John Ferguson McLennan

THE PATRIARCHAL THEORY

Elibron Classics
www.elibron.com

Elibron Classics series.

ISBN 0-543-92602-8 (paperback)
ISBN 0-543-92601-X (hardcover)

This Elibron Classics Replica Edition is an unabridged facsimile of the edition published in 1885 by Macmillan and Co., London.

THE PATRIARCHAL THEORY.

THE

PATRIARCHAL THEORY

BASED ON THE PAPERS OF

THE LATE

JOHN FERGUSON McLENNAN

Edited and Completed

BY

DONALD McLENNAN M.A.

OF THE INNER TEMPLE, BARRISTER-AT-LAW.

LONDON

MACMILLAN AND CO.

1885

CHARLES DICKENS AND EVANS,
CRYSTAL PALACE PRESS.

PREFACE.

In the last years of his life my brother was engaged, amidst many difficulties from failing health and prolonged absences from England, in following up the lines of inquiry first opened by his book on Primitive Marriage and his essay in the *Fortnightly Review* on the Worship of Animals and Plants.*

The views as to the early structure and subsequent movements of human society embodied in these essays were, on the whole, confirmed and enlarged by further study;† the range of available evidence was gradually extended by a systematic survey of all authentic accounts of primitive peoples in the various quarters of the globe; at the same time new problems arose, and new points of view suggested themselves as likely to furnish their satis-

* "The Worship of Animals and Plants," *Fortnightly Review*, 1869–70.

† Some of his views were modified more or less, and it will appear from this work that one was abandoned—the view expressed in *Primitive Marriage* that Agnation was, at a certain stage, generally prevalent, in stating which it may be believed that he yielded somewhat to the authority of Sir Henry Maine.

factory solution, and ultimately it seemed no hopeless task, if only health and strength had revived, to undertake a general work on the structure of the earliest human societies. In particular, he felt that he was able to give a much more consistent and intelligible view of the condition of rude or undeveloped communities than anything that had previously been offered to the public. It was a part of his design to set forth a theory of the Origin of Exogamy, and to gather together the facts, very numerous and falling into several classes, by which that theory could be supported.

But the inquiry into Exogamy and into the conditions of rude society in which it must be supposed to have originated led to another generalisation. As the theory of the Origin of Exogamy took shape, and the facts connected with it reduced themselves to form in his mind, the conclusion was reached that the system conveniently called "Totemism"—from which his essay on the Worship of Animals and Plants took its departure—must have been established in rude societies prior to the origin of Exogamy. This carried back the origin of Totemism to a state of man in which no idea of incest existed. From that condition my brother hoped to be able to trace the progress of Totemism—necessarily a progress upwards—in connection with kinship and with Exogamy. It may here be said that he had for a time a hypothesis as to the origin of Totemism, but that he afterwards came to see that there were conclusive reasons against it. At last, as far as I know, he had none—which should be easily intelligible to anyone who knows the subject and knows what, on his view, was involved in Totemism. To show its prevalence, to establish some leading points in its history, to exhibit it in connection with kinship and with Exogamy, and to make out its connection with worship appeared to him to be the matters primarily important.

It may be said that evidence of Totemism associated with Exogamy was generally found in all rude societies acknowledging kinship through women only; that the same association was found also, and almost as generally, in rude societies which know kinship through males; while his original essay had tended to show that the worship of plants and animals in more advanced societies acknowledging kinship through males was lineally descended from Totemism.

The general conclusion from these and allied facts taken as a whole, appeared to be that it was possible to demonstrate that, Totemism preceding Exogamy, the latter must have arisen in societies acknowledging no kinship save through women; that all other facts bearing on rude society may be interpreted as evidence of a gradual progress from the condition of which Totemism and female kinship are the mark; and that thus it was possible to exhibit the history of human society as that of an evolution moving with very various rapidity among different populations, but always beginning with a condition in which the idea of incest did not exist, and always tending upwards from that condition.

With regard to the part to be played in this research by the theory of Exogamy, one other point may be noticed. It was found that Exogamy had an extension so great, and of such a kind, as to imply that it arose from the operation of *general* causes not limited to the circumstances of this or that particular population. Now Exogamy is found alike in rude societies with kinship through males, and in the still ruder societies which have kinship only through females. It is perhaps scarcely a legitimate hypothesis to suppose that it had two separate origins for these two kinds of societies. And the presumption seems to be that it began with the ruder of the two; in other words, that it began before the recognition of male kinship. At any rate, as has been said already, the explanation of its origin which my brother designed to set forth was in accordance with this presumption. And here the argument from Totemism came in to enforce and justify the conclusion by making it probable that every society in which Exogamy has been found, must at one time have known kinship through females only. Now the general cause from which his theory derived the origin of

Exogamy was a scarcity of women. And the point to be noticed is that in this connection my brother had made large collections as to the prevalence of infanticide and kindred practices, classified in relation to the systems of kinship with which they are associated.

These indications will suffice to show that the proposed research was of a very extensive and far-reaching kind, and involved the use of a very large apparatus of evidence. This being so, my brother proposed to prepare the way for his larger work by first issuing a critical essay by which he hoped to clear out of the way a body of opinion, the prevalence of which seemed to oppose an obstacle to the proper appreciation of his constructive argument.

From the time of Plato downwards, theories of human society have been current in which the family living under the headship of a father is accepted as the ultimate social unit. These theories have taken various shapes, but in his opinion the most important, as well as the most influential, shape to be taken account of is that represented in the works of Sir Henry Maine. In the hands of the more prominent amongst its older advocates within the period of modern thought the Patriarchal Theory, as it is called, was mainly a theory of the source of sovereignty, and in this aspect it had gradually ceased to attract attention. With Sir Henry Maine, on the other hand, the theory becomes a theory of the origin of society, or at least of the earliest stage of society in which Comparative Jurisprudence is called upon to take interest. And at the same time the theory is expressly based on a comparative study of early societies, so that it comes into direct conflict with every theory of the origin of society which does not accept the family as the primitive unit.

It was necessary, therefore, for my brother to take notice of this theory, and to do so on the scale which the intrinsic importance of the question demanded. To make such a discussion a mere incident in a large and complicated constructive work would not have tended to convenience or clearness, and the only alternative was to make it the subject of a separate essay. This plan had the disadvantage that such an essay could not but be polemical in form, a thing which has always a somewhat ungracious look; but it was felt that the enormous prestige which the Patriarchal Theory has gained through Sir Henry Maine's advocacy made it impossible to separate the argument from the form which he has impressed on it. In English-speaking countries, at least, this is the one form in which the theory is current, and no discussion of it would have been useful which did not closely follow the statements of the author of *Ancient Law*, a book which, for more than twenty years, has profoundly influenced the whole teaching of Jurisprudence in our country. An independent thinker who has arrived at conclusions fundamentally at variance with those set forth in so influential a work cannot avoid bringing the points in dispute to a direct issue.

The proposed criticism of the Patriarchal Theory was first thought of in 1879, but the execution of the project was deferred by increasing illness. Some notes, however, were made from time to time, and in the winter of 1880 I began, at my brother's request, actually to write the book in co-operation with him, and we finished together a draft of the first six chapters of the present essay, and also of the ninth, the substance of which had for the most part been published in his *Studies in Ancient History.* His last illness cut short the work at this point, and what remains I have had to work out for myself, with the aid of some fragmentary notes, mostly relative to Agnation, expressly prepared for this work, and—for the argument presently to be spoken of—of the hints that could be derived from the collections—also fragmentary—formed for his larger project.

Fortunately the notes referred to included a short paper containing what appeared to me to be invaluable suggestions for what I venture to think the most important part of the whole discussion, namely, the origin of Agnation. This paper (besides that it was pointed out that, on Sir Henry Maine's statement, Agnation might have been expected to fall with Patria Potestas, unless something could be indicated capable of making it survive; and that at Rome Patria Potestas was the longer-lived of the two) gave me the suggestion that Agnation was originally the gentile relationship, the suggestion that this could be supported by the analogous operation of female kinship, the observation that Agnation is none the less an exceptional phenomenon, and the indication of the retarding influences which may have prevented its more general occurrence. To bring out the full scope of these important hints, I found it necessary to build up as best I could a somewhat elaborate argument, in which the original scheme of the work was necessarily transcended, so that the latter part of the book is more constructive and less purely critical than had been at first designed. Another reason for this was the necessity for revising the whole discussion in the light of Sir Henry Maine's latest work, which made it needless to dwell at length on several parts of the original plan, while on the other hand it rendered it inevitable to go fully into the whole question of the Levirate and of the family custom of the Hindoos. For this part of the book of course I alone am responsible.

I should add, perhaps, that I had hoped to bring the book out much sooner. But the state of my health has compelled me to work slowly. And I had to make some laborious researches.

It remains for me to express my gratitude to Professor W. R. Smith, of Cambridge, for his kindness in reading a proof of this work, and for his readiness on all occasions to give me help and save me trouble. Some valuable notes which he has contributed to the book are acknowledged at the proper places.

D. McLENNAN.

TEMPLE, *November* 14*th*, 1884.

CONTENTS.

CHAPTER IX.

CHAPTER X.

CHAPTER XI.

CHAPTER XII.

CHAPTER XIII.

CHAPTER XIV.

CHAPTER XV.

CHAPTER XVI.

CHAPTER XVII.

CHAPTER XVIII.

CHAPTER XIX.

THE PATRIARCHAL THEORY.

CHAPTER I.

THE PATRIARCHAL THEORY.

The Patriarchal Theory, stated in its simplest form, represents society as the enlargement of the family, and the family as a group composed at first of a man and his wife and children. When the children, and afterwards their children and more distant descendants, married, many such groups would be formed round the original family, and all of these would acknowledge the authority of the First Father, as chief or patriarch, as long as he lived. However large the body of descendants might become, they would constitute but one family, of which the First Father would be the natural head. On his death his descendants would naturally divide into as many families as he had sons with offspring. Each of these would resemble the original group absolutely—would be, that is, a collection of persons connected by common descent, living under the authority of their common progenitor. What had happened on the death of the First Father would happen thereafter to every family on the death of its head; it would be resolved into a series of families, each under the headship of that son of the deceased from whom it was derived.

Should all the families descended from the same First Father continue to hold together, they would in time form a very large group—in short, a tribe; and thus the Theory explains the genesis of tribes. It can also be made to furnish a partial explanation of the formation of gentes, clans, or houses, within the tribe. These would begin to be formed as soon as some principle of succession or election to the chieftainship preserved families from dissolving on every death of a chief. Or they might arise through the families descended from some eminent progenitor distinguishing themselves from the other families of the tribe by a name indicative of such descent and of a special connection between them. The Theory, moreover, explains the genesis of nations. As there would be the nucleus of a new tribe whenever a man descended from the First Father separated, with his wife and children, from the main body of his kindred, and settled in a new district, in the course of generations—after many such separations had taken place—the descendants of the same First Father might constitute many tribes and be the population of a large country. The tribes, being united by ties of blood, would readily act together for common purposes. By-and-by they would establish some form of central government to facilitate such action. Then they would have become a nation.

It has been usual to cite the history of Israel as an illustration of this theory as stated above. Each of the twelve tribes consisted of descendants of a son or grandson of Jacob, the First Father; and their union constituted the Jewish nation. In Genesis, moreover, the population of the world is represented as composed of tribes and nations deduced from the sons of Noah —Shem, Ham, and Japhet.

The Patriarchal Theory, so stated, was most simple, and agreeable to current prejudices. It used to be generally accepted as palpably true, like the fact of the sun moving daily round the earth. No one thought of proving it, and but few of seriously doubting it. But there were facts against it, and at length some of them were noticed. Its explanation of the origin of gentes, or clans, within the tribe was obviously insufficient. Tribes like the Roman Tribe, consisting of a number of different stocks—consisting of clans, each of which differed in blood or origin from the others—have been exceedingly common; and of their composition the theory could give no account at all. What it could prove was that the clans composing a tribe must always and everywhere be all of the same stock —viz., that of the First Father of the tribe. The insufficiency of this theory, however, such was its hold upon the world, suggested, in the first instance, not any doubt of its truth as a whole, but the need of making some modification of it, or of finding some means of supplementing it. And in a modified form, and supplemented to some extent, "Comparative Jurisprudence" was promising to establish it for ever, beyond the reach of doubt, at the time when—research setting ancient facts in new points of view—it first came seriously into question.

As thus modified and supplemented the Patriarchal Theory holds its ground among us. And it is as thus modified and supplemented that it is here to be the subject of criticism. Its insufficiency in its old form seems to be admitted. On the other hand, as now presented, it lacks the simplicity which was formerly its best recommendation.

The ingenious and learned author of *Ancient Law*, indeed, in restating the Patriarchal Theory, ascribed to it features which made it very different from the view which had held possession of the world, and which might reasonably have hindered its acceptance. The Patriarchal Family, as he conceived of it—complex, artificial, strange—does not look as if it could belong to the earliest history of man.

It is not merely a group of descendants with the First Father at their head. It is a group of persons living under a Patriarch who has over them despotic power, and can sell any of them, or put him to death; and they are held to be related to him, and to one another, not so much because of their being of his blood as because of their common subjection to his power. As to the composition of this group, the Patriarch having children and other descendants, it includes children and other descendants of his, and, in theory, it is made up of his descendants. But the stranger whom he adopts—and he practises adoption largely—is in every respect as a son. On the other hand, the tie of blood counts for so little that a son who leaves him ceases to be a relation of him and his family ; and a daughter, when she marries, becomes as a stranger to them ; all descendants through daughters being also excluded from relationship to the Patriarch and his house. The blood-connection between the family and one-half of its kinsfolk is always ignored, while a firm bond of relationship connects people who have no blood in common. Primitive as it is, this family includes slaves as well as children and adopted persons. The slave, however, is not within the relationship, because he is to remain a slave for ever. As to the powers supposed to have been wielded by the Patriarch over the members of the Primeval Family, they are that assortment of powers which in Rome was called Patria Potestas; while the system of relationship is what was conceived of as the simplest form of the Roman relationship, *Agnation*—that system which, in a more developed form, severed from all rights of succession to family property, and even from the family, every person connected with its head through women only.

It is mainly on a proof of the universal prevalence in the earliest times of the Roman institutions of Patria Potestas and Agnation that this novel view of the Patriarchal Family is founded. The Primeval Family of the theory differs, however, from the ancient Roman family by which it was suggested, in respect of the use it made of adoption. It was freely enlarged by the practice of adoption; whereas in Rome adoption was originally—as the Will (to which it was prior in origin) also was—merely an expedient to enable a man who had no child, and no prospect of having one, to choose a young man, usually a near relative, to be to him as a son—to continue his family, to keep up his sacra, and, on the other hand, to inherit his property.

The following extracts from *Ancient Law* will be enough to satisfy the reader of the accuracy of the account just given of the view propounded in that work:

1. *Effect of the Evidence.*—"The effect of the evidence derived from Comparative Jurisprudence is to establish that view of the primeval condition of the human race which is known as the Patriarchal Theory."—*Ancient Law*, p. 122.*

2. *The Primeval Family.*—"The eldest male parent is absolutely supreme in his household. His dominion extends to life and death, and is as unqualified over his children and their houses as over his slaves; indeed, the relations of sonship and serfdom appear to differ in little beyond the higher capacity which the child in blood possesses of becoming one day the head of a family himself."—*Ibid.*, p. 123.

"The Family is the type of an archaic society in all the modifications which it was capable of assuming. . . . We must look on the family as constantly enlarged by the adoption of strangers within its circle, and we must try to regard the fiction of adoption as so closely simulating the reality of kinship, that neither law nor

* The paging is the same in all the editions, as far as this subject is concerned. The writer has before him the first edition (1861), and the seventh (1878). London: Murray.

opinion makes the slightest difference between a real and an adoptive connexion. On the other hand, the persons theoretically amalgamated into a family by their common descent are practically held together by common obedience to their highest living ascendant, the father, grandfather, or great-grandfather. The patriarchal authority of a chieftain is as necessary an ingredient in the notion of the family group as the fact (or assumed fact) of its having sprung from his loins; and hence we must understand that if there be any persons who, however truly included in the brotherhood by virtue of their blood relationship, have nevertheless *de facto* withdrawn themselves from the empire of its ruler, they are always, in the beginnings of law, considered as lost to the family. It is this patriarchal aggregate—the modern family thus cut down on one side and extended on the other—which meets us on the threshold of primitive jurisprudence."—*Ibid.*, pp. 133–4.

3. *Agnation.*—"Agnatic relationship is in truth the connexion existing between the members of the Family conceived as it was in the most ancient times."—*Ibid.*, p. 147.

"The foundation of Agnation is not the marriage of father and mother, but the authority of the Father. . . . In truth, in the primitive view, Relationship is exactly limited by Patria Potestas. Where the Potestas begins, Kinship begins, and therefore adoptive relatives are among the kindred. Where the Potestas ends, Kinship ends, so that a son emancipated by his father loses all rights of Agnation. And here we have the reason why the descendants of females are outside the limits of archaic kinship."—*Ibid.*, p. 149.

"Cognates are all those persons who can trace their blood to a single ancestor and ancestress." . . . Agnates "are all the cognates who trace their connexion exclusively through males," with or without the addition of persons brought into the family "by the artificial extension of its boundaries," as, *e.g.*, by adoption. —*Ibid.*, pp. 147–8.

"It is obvious that the organisation of primitive societies would have been confounded if men had called themselves relatives of their mothers' relatives."—*Ibid.*, p. 149.

4. *Universal Prevalence in Primeval Times of Patria Potestas and Agnation.*—"The Patria Potestas, in its normal state, has not been a durable institution. The proof of its former universality is therefore incomplete as long as we consider it by itself."—*Ibid.*, p. 146.

"Hence comes the interest of Agnation for the inquirer into the history of jurisprudence. The Powers [Patria Potestas] themselves are discernible in comparatively few monuments of ancient law; but Agnatic relationship, which implies their former existence, is discoverable almost everywhere."—*Ibid.*, p. 150.

The view disclosed in the passages quoted—and it is stated without reservations or ambiguities—is, in effect, that the family, much as it existed among the Romans within the historical period, was primeval and universal; and it is a proof of this that is put forward as establishing the Patriarchal Theory. It should be noticed that Agnation is declared to be the only form of relationship consistent with the structure of primeval society, and therefore the only form of relationship known in the most ancient times. Cognation or natural relationship, had it been acknowledged, would have led to the organisation of primitive societies being *confounded*, and therefore in primitive times it could not be recognised.

The Patriarchal Theory, in its new form, could not, any more than in its old form, account for the union, otherwise than as rulers and ruled, of different stocks in the same society. The author of *Ancient Law*, however, thought that an explanation of the actual heterogeneity of societies which, according to the theory, should have been homogeneous, was to be found in the use in early times of a fiction analogous to adoption and having a similar effect. His view is propounded in vague terms, but the following passages will show what it is, so far as it is defined:

The Genesis of Society.—"In most of the Greek states and in Rome, there long remained the vestiges of an ascending series of groups out of which the State was at first constituted. The Family, House, and Tribe of the Romans may be taken as the type of them, and they are so described to us that we can scarcely help conceiving them as a system of concentric circles which have gradually expanded from the same point. The elementary group is the Family, connected by common subjection to the highest male ascendant. The aggregation of Families forms the Gens or House. The aggregation of Houses makes the Tribe. The aggregation of Tribes constitutes the Commonwealth."—*Ancient Law*, p. 128.

A Difficulty. The Heterogeneousness of Early Communities.—" It may be affirmed of early commonwealths that their citizens considered all the groups in which they claimed membership to be founded on common lineage. . . . And yet we find that along with this belief, or if we may use the word, this theory, each community preserved records or traditions which distinctly showed that the fundamental assertion was false. . . . The composition of the state, uniformly assumed to be natural, was nevertheless known to be in a great measure artificial."—*Ibid.*, pp. 129, 130.

How the Difficulty is dealt with.—" The earliest and most extensively employed of legal fictions was that which permitted family relations to be created artificially, and there is none to which I conceive mankind to be more deeply indebted. If it had never existed, I do not see how any one of the primitive groups, whatever were their nature, could have absorbed another, or on what terms any two of them could have combined, except those of absolute superiority on one side and absolute subjection on the other. No doubt, when with our modern ideas we contemplate the union of independent communities, we can suggest a hundred modes of carrying it out, the simplest of all being that the individuals comprised in the coalescing groups shall vote or act together according to local propinquity; but the idea that a number of persons should exercise rights in common, simply because they happened to live within the same topographical limits, was utterly strange and monstrous to primitive antiquity. The expedient which in those times commanded favour was that the incoming population should *feign themselves* to be deduced from the same stock as the people on whom they were engrafted; and it is precisely the good faith of this fiction, and the closeness with which it seemed to imitate reality, that we cannot now hope to understand."—*Ibid.*, pp. 130, 131.

That each of the larger groups was an "aggregation" of several groups of the order immediately lower; that the practice of creating family relations artificially somehow helped on, or may have helped on, the aggregation of groups that were not of the same origin; and that the expedient actually employed was "that the incoming population should feign themselves to be deduced from the same stock as the people on whom they were engrafted" appears to be the view suggested. It is not within the purpose of the present work to consider it closely; but the following observations may be offered:

1st. That peoples not of the same origin who had become united, in after times considered themselves of the same stock, is unquestionable.

But did they use the pretence, known to them to be false, that they were of the same stock to bring about their combination? or was this only a hypothesis by which the fact of their living in combination was ultimately accounted for?

2nd. The populations which are said to have combined on the pretence that they were of the same descent, remained distinct from one another after their union, each retaining its own name. The fiction employed, therefore, had not the same effect as adoption. It did not merge the one population in the other so that no trace of heterogeneity was left. By adoption, on the other hand, the adopted person became merged in his new family, became a part of it, as if he had been born in it, and nothing survived to show that a stranger had been brought in.

3rd. Adoption itself can account for no appearance of heterogeneity in Gens or Tribe, because, as has just been said, it left no mark upon the family, nothing to indicate that it had occurred. And there is another reason why no such effect can be attributed to it. In the best-known cases, the person who could be adopted was a person of the adopter's blood.

"How," says Sir A. C. Lyell,* "does it come to pass that in those primitive societies which assume as their basis a common descent from one original stock, one so constantly finds traces of alien descent? How came a variety of alien groups to coalesce into a local tribe? The fiction of male adoption is suggested as an answer, but such adoption from alien stocks is quite unknown throughout India, where the adoption of a son is always

* "Formation of Indian Clans and Castes," *Fortnightly Review*, January, 1877. Sir A. Lyell's evidence as to the practice of adoption among modern Hindoos is presumably good. On the value of his speculation as to the origin of Clans and Castes it is unnecessary here to offer any opinion.

made within the circle of affinity, ordinarily from the nearest kindred." Adoption, in fact, does not in India bring any alien blood into a family.

In Rome, too, the person brought into a family by adoption usually belonged to the same Gens, that is stock, as the adopter, and he took his place in the adopter's family just as if he had been born in it.

Even if, at Rome, the adopted person belonged to a different Gens from the adopter, he was, at any rate, separated absolutely from the Gens of his birth and its sacra, and introduced to the sacra of the adopter's Gens and family.* He bore both the Gentile and the family name of the adopter. Thus the adoption left no permanent trace—there remained nothing to show that a stranger had come into the family.

Neither in India nor in Rome, then, could adoption account for the presence of Gentes of different stocks within the same local tribe, or for the appearance of families apparently of different stocks within the same Gens. So far of adoption in advanced societies.

To pass to the primitive races (so called), we find that among them, too, a person adopted takes his place in the family of adoption as if he had been born in it. When a captive taken in war is adopted to fill the place of a person recently lost, he usually takes the very name and place in the family of the deceased. In a widely extended class of cases he takes the Gentile name of the adoptive mother. In the cases more

* It is extremely doubtful whether this case ever occurred in Rome, though some Roman lawyers say it did.

immediately comparable with those of Rome and India —cases in which kinship is through males and children are of the stock of their father—he takes the (totem) Gentile name of the adoptive father, and is as much under the bond of blood involved in that as if he had by birth belonged to the father's stock. Among primitive races, then, just as among more advanced races, adoption leaves no trace of a stranger having been brought into the stock. It produces, that is, no permanent appearance of heterogeneity.

How far familiarity with a fiction which produced the *appearance* of kinship could dispose ancient groups everywhere, or nearly everywhere, to employ in uniting with one another a fiction which did not produce the appearance of kinship—nor, so far as appears, any effect whatever—is a matter on which it would be idle to speculate.

It cannot now be too much to say, however, that the view propounded by way of supplement to the Patriarchal Theory is itself so far wrapped in obscurity, and so much open to doubt that the unfitness of the theory by itself to explain the growth of society ought to continue to tell against it. Here it will only be made one of many reasons why it should be held indispensable that the evidence for the Patriarchal Theory should be full, clear, and strong.

CHAPTER II.

SIR HENRY MAINE'S LATER WRITINGS.

SINCE the publication of *Ancient Law* much has been written which has tended to raise doubts as to the soundness of the Patriarchal Theory; and, in the later writings of the author of that work, it is evident enough that he has at times been somewhat troubled about the validity of his early impressions. The Patriarchal Family of his theory has seemed to himself, considering it afresh, a strange and inexplicable institution; and he could not but see that, if the reports of observers are to be trusted, there have been many bodies of men among whom it has been unknown. Excepting in a single passage in his latest work, however,* reconsideration has never carried him beyond the admission of a bare possibility —a possibility too faint, apparently, to be worth thinking seriously about, and which, at any rate, comparative jurisprudence need not concern itself with—that the Patriarchal Family was not a primary social fact. It has left him able to state as confidently as he did at first—the passage just referred to making no exception

* *Early Law and Custom*, pp. 286–288.

to this—that, among all the more important of early tribes, among all tribes that are worth any reasonable person's notice, the Patriarchal Family, as he has described it, with despotic power in the father, and Agnation as the only bond of relationship, is to be met with at the beginning of history, either actually existing, or plainly traceable by its incidents and the marks it has left upon law and custom. The passages subjoined are from his later works, and it will be found they bear out what has just been said:

1. *The Strangeness of the Patriarchal Family.*—"The Patriarchal Family is not a simple but a highly complex group, and there is nothing in the superficial passions, habits, or tendencies of human nature which at all sufficiently accounts for it. If it is really to be accepted as a primary social fact, the explanation assuredly lies among the secrets and mysteries of our nature, not in any characteristics which are on the surface. Again, under its best ascertained forms, the Family Group is in a high degree artificially constituted, since it is freely recruited by the adoption of strangers. All this justifies the hesitation which leads to further inquiry."—*Village Communities in the East and West*, pp. 15, 16.

The Patriarchal Theory. Description of the Patriarchal Family.—"The two societies, Roman and Hindoo, which I take up for examination . . . are seen to be formed at what for practical purposes is the earliest stage of their history, by the multiplication of a particular unit or group, the Patriarchal Family.

There has been much speculation of late among writers belonging to the school of so-called prehistoric inquiry, as to the place in the history of human society to which this peculiar group, the Patriarchal Family, is entitled. Whether, however, it has existed universally from all time—whether it has existed from all time only in certain races—or whether, in the races among whose institutions it appears, it has been formed by slow and gradual development—it has everywhere, where we find it, the same character and composition. The group consists of animate and inanimate property, of wife, children, slaves, land and goods, all held together by subjection to the despotic authority of the eldest male of the eldest ascending line, the father, grandfather, or even more remote ancestor. The force which binds the group together is Power. A child adopted into the Patriarchal Family belongs to it as perfectly as the child naturally born into it, and a child who severs his connection with it is lost to it altogether. All the larger groups which make up the primitive societies in which the Patriarchal Family occurs, are seen to be multiplications of it, and to be, in fact, themselves more or less formed on its model."—*The Early History of Institutions*, pp. 310, 311.

Prevalence of the Patriarchal Family.—"Among the Aryan sub-races, the Hindoos may be as confidently asserted as the Romans to have had their society organised as a collection of patriarchally-governed families."—*Ibid.*, p. 323.

"My suggestion is that the key to the Irish distribution of the Family, as to so many other things in ancient law, must be sought in the Patria Potestas."—*Ibid.*, p. 217.

"I think I can assign some at least plausible reasons for believing that this perplexing fourfold division of the Celtic family is . . . a monument of that Power of the Father which is the first and greatest landmark in the course of legal history."—*Ibid.*, p. 216.

"Meantime, let me say something on the transmutations which Patriarchal Power is observed, as a fact, to undergo in the assemblages of men held together by kinship which are still found making a part of Aryan communities. The Joint Undivided Family, wherever its beginning is seen in such communities, springs universally out of the Patriarchal Family, a group of natural or adoptive descendants held together by subjection to the eldest living ascendant, father, grandfather, or great-grandfather. Whatever be the formal prescriptions of the law, the head of such a group is always in practice despotic, and he is the object of a respect, if not always of an affection, which is probably seated deeper than any positive institution."—*Ibid.*, pp. 115, 116.

"There can be no reasonable doubt that the House Community of the South Slavonians is the Roman Gens, the Hellenic γένος, the Celtic Sept, the Teutonic Kin. It is also the Joint Family of the Hindoos, which is itself a living though an extremely perishable institution."—"South Slavonians and Rajpoots," *Nineteenth Century*, Dec., 1877.

In *Early Law and Custom*, published in 1883, this passage reappears with a modification. "There can be no reasonable doubt," it is there said, "that the House Community of the South Slavonians corresponds to one or other of the larger Roman groups"—that is, either to the Gens or the body of Agnates.

"I have, however, no doubt myself, from a variety of indications, that these families [' natural families,' consisting of the descendants of an ancestor still alive] are, to employ a convenient term, patriarchal families despotically governed by the eldest ascendant. . . . The South Slavonians, like the Romans, maintain a clear distinction between Agnatic and Cognatic relationship, which they term respectively kinship through the great blood and kinship through the little. Thus a group of men connected with a common ancestor through male descents (natural or adoptive) exclusively, are kinsmen of the great blood; they are kinsmen of the little blood when they include also the descendants of female relatives. Now the recognition of Agnatic relationship is good evidence that patriarchal power either exists or has once existed in a community; there may have been paternal power where there is no Agnation, but where there is Agnation there must almost certainly have been paternal power."—"South Slavonians and Rajpoots," *Nineteenth Century*, 1877.

This passage reappears, with a slight modification in the first sentence quoted, in *Early Law and Custom*, pp. 243, 244. It is there said that the families are, "as a rule," despotically governed by the eldest ascendant.

"The most recent researches into the primitive history of society point to the conclusion that the earliest tie which knitted men together in communities was consanguinity or kinship. . . . . It was regarded as an actual bond of union, and in no respect as a sentimental one. . . . . If a man was not of kin to another there was nothing between them. He was an enemy to be slain, or spoiled, or hated. . . . . The tribes of men with which the student of jurisprudence is concerned are exclusively those belonging to the races now universally classed, on the ground of linguistic affinities, as Aryan and Semitic. Besides these he has at most to take into account that portion of the outlying mass of mankind which has lately been called Uralian—the Turks, Hungarians, and Finns. The characteristic of all these races, when in the tribal state, is that the tribes themselves, and all sub-divisions of them, are conceived by the men who compose them as descended from a single male ancestor. Such communities see the family group with which they are familiar to be made up of the descendants of a single living man, and of his wife or wives; and perhaps they are accustomed to that larger group, formed of the descendants of a single recently deceased ancestor, which still survives in India as a compact assemblage of blood relatives, though it is only known to us through the traces it has left in our Tables of Inheritance. The mode of constituting groups of kinsmen which they see proceeding before their eyes they believe to be identical with the process by which the community itself was formed. Thus the theoretical assumption is that all the tribesmen are descended from some common ancestor, whose descendants have formed sub-groups, which again have branched off into others, till the smallest group of all, the existing Family, is reached."—*The Early History of Institutions*, pp. 64–66.

It will be seen that, in these passages, the Patriarchal Family of *Ancient Law* appears unchanged; that fresh examples of Patria Potestas are noted; and that the use of Agnation as furnishing a clue to the previous existence of Patria Potestas, and therefore of the Patriarchal Family, is still insisted upon. The Patriarchal Family—or sure indications of it—is found among Romans, Greeks, Hindoos, Celts, Teutons, and Slavonians; while the Patria Potestas is seen on all hands affecting ancient law, and is, in fact, "the first and greatest landmark in the course of legal history." The passage last quoted is an allegation that the Patriarchal Theory has been held by all Aryan, Semitic, and Uralian tribes, and an attempt to show how they came to believe in it.

It should be added that, so far as the matters under consideration go, *Ancient Law* itself has remained absolutely unchanged since 1861; and, as regards the "proof," from comparative jurisprudence, of the truth of the Patriarchal Theory therein contained, that Sir Henry Maine referred to it in his *Village Com-*

*munities** as a proof to which he adhered in toto. "I need not here repeat to you," he says, "the proof which I have attempted to give elsewhere."

The passage in *Early Law and Custom* referred to in the first paragraph of this chapter, makes in some respects a fresh departure from the author's other writings. It occurs in a note appended to one of the essays contained in that volume, and is as follows:

"The 'Agnatic' Gentile groups, consisting of all the descendants, through males, of a common male ancestor, began to exist in every association of men and women which held together for more than a single generation. They existed because they existed in nature. Similarly the group consisting of the descendants, through women, of a single ancestress still survives, and its outline may still be marked out, if it be worth anybody's while to trace it. What was new at a certain stage of the history of all or a portion of the human race, must have occurred, not in connection with the Gens, but in connection with the Family. There was always one male parent of each child born, but prevalent habits prevented his being individualised in the mind. At some point of time, some change of surrounding facts enabled paternity, which had always existed, to be mentally contemplated; and further, as a consequence of its recognition, enabled the kinship flowing from common paternity to be mentally contemplated also. As to the new facts which led to this recognition, all that, in my opinion, can be said of them

* *Village Communities in East and West*, p. 15. London, 1871.

is that they must have been such as again to give free play to an over-mastering emotional force. Believing, as I do, that when Paternity reappeared, it reappeared in association with Power and Protection, I require no explanation of the fact that the kinship then recognised was kinship through male descents only."—*Early Law and Custom*, pp. 287, 288.

Here we seem to have a period during which there is an obscuration of paternity—an inability to individualise the father—admitted for "all or a portion of the human race." But the Patriarchal Family emerges at the close of it. "Power and Protection" in a father may no doubt mean less than Patria Potestas; but *Early Law and Custom*—and, indeed, the essay in it to which the note in which this passage occurs is appended—has so much to say of Patria Potestas, and of the father's despotic power, that it is not to be supposed that less is meant.

CHAPTER III.

THE PATRIARCHAL FAMILY. THE AMOUNT OF PROOF OF IT THAT MAY SUFFICE.

In a passage already quoted, Sir Henry Maine has told us that "the effect of the evidence derived from Comparative Jurisprudence" is to establish the Patriarchal Theory, as he has stated it. We are going to give some reasons why it should be deemed necessary that the evidence for the Patriarchal Family of his theory, with its incidents of Patria Potestas and Agnation, should be exceedingly clear and strong.

1. A sound theory of the origin of society should explain at least the leading facts connected with the growth of societies, easily and effectually; and this, as is frankly admitted by Sir Henry Maine, the Patriarchal Theory cannot do. That fictions, or other such expedients, should play some part in the forming of societies would be in no way surprising. But Sir Henry Maine has to throw the whole work of accounting for the commonest and most important facts of social organisation upon a fiction, and that a fiction of his own supposing—his Patriarchal Theory being

a mere stumbling-block to him. And even as to this fiction he has to admit that its having been used with good faith is "what we cannot hope to understand"—that its having been employed, that is, if it was employed, is a thing utterly surprising and unintelligible. Surely this failure of the theory raises a presumption, or, at any rate, a feeling requiring to be overborne by evidence, that the family of the theory cannot have been the factor in social growth it is said to have been. On this account alone it would be proper to call for good evidence of its prevalence, and for good reasons for thinking that, where found, it was primordial.

2. A passage that has been quoted shows that Sir Henry Maine himself thinks it only natural that people should doubt whether the family of his theory is to be accepted as a primary social fact. It is not a simple but a highly complex group, he admits; "there is nothing in the superficial passions, habits, or tendencies of human nature which at all sufficiently accounts for it;" if it be a primary social fact, "the explanation assuredly lies among the secrets and mysteries of our nature."

And, in truth, the Family held together by Power, with blood-relationship recognised in it only to be ignored—no relationship at all through women acknowledged, no relationship through males acknowledged except in males subject to the father's Power, and between those subject to that Power, a relationship equally close whether they are related by blood or not—the Power, too, extending to life and death and sale, and grown-up sons meekly submitting to it—propounded to us as the first form of the family, might well be deemed—apart from evidence—a mere fantastic imagination. Strangely complex as it is, the evidence should be good indeed on which it is accepted as having been primordial and universal.

As to the complexity of this family, no doubt too much importance may be attached to the idea that institutions in their beginnings are usually simple, and become complex as they grow old. It is an idea borne out, however, by many facts from many fields of nature. No one could believe in the Ornithorynchus as a germinal type of animal life. But the family of Sir Henry Maine's theory is almost as curious a complex of types as is the Ornithorynchus. Its head is head partly as being the begetter, and partly as being the owner, of its members; so that the cementing principle is neither kinship nor property, but a jumble of the two. Kinship is not excluded, for *in theory*—that is, partly in fact, and partly by a fiction—the family is made up of the father's descendants, and he is the representative of the family, not its owner; and, on the other hand, he has over the members of it, and over all that pertains to it, an uncontrolled and unlimited power of disposal. Then it may almost be said to be based upon fictions. By a fiction, the wife, the mother of the family so far as its members are begotten, is not the wife of her lord, but his daughter, and the sister of her own children. The children begotten are, in fact, *property* of the father, and, by a fiction, cease to be his children if he sells them. By a further fiction, additional children, who become in the full sense members of the family, may be acquired by him by purchase, or a pretence of purchase, and be to him even as sons and daughters of his blood. Can anyone believe, excepting for convincing reasons, that such a group as this was elementary and primordial? Is it not to be presumed that, if found at all, it must have a history through which its artificialities can be explained? It would be as easy to believe in Minerva springing full-armed from the head of Jove as to believe, except under the constraint of evidence, in such a family as rudimentary. To return to our point, surely the evidence to compel belief ought to be exceedingly full and good. Perhaps this has now been sufficiently made out, but there is something more which it would be improper altogether to overlook.

3. There are many rude societies now existing in which the family is radically different from the Patriarchal Family; and there is a great mass of evidence which goes to show that an earlier family system, founded solely on the recognition of blood-ties, though of those only which men first learned to acknowledge, everywhere preceded that family system, which, with strange and peculiar incidents superadded, and blood-relationship almost ignored, appears in Sir Henry Maine's theory. It can be shown how this other family system, founded on the recognition of kinship through women, and through women only, would naturally give

place to a family system resembling (to speak loosely) that of the theory—to a family system in which the father is head of the family, and the children are counted of his blood or stock. And, indeed, the transition from the one system to the other can be clearly illustrated by actual cases, and many of the intermediate stages mapped out. The actual heterogeneousness of societies, too—not that the agency of minor causes in producing that need be called in question—is simply and effectually accounted for by means of the kinship acknowledged in this form of the family. It has appeared that the Patriarchal Theory fails to explain the heterogeneousness of societies; and it seems impossible to travel onwards from the Patriarchal Family depicted in *Ancient Law* to the family founded on the recognition of kinship through women only, and show how the one can have changed into the other.

Comparing these two forms of the family, indeed—assuming both to have existed—there cannot be a question which is the earlier of the two, or which is, so to speak, the more natural. In the one, blood-relationship, struggling into recognition, is found embodied in a system than which there can have been no earlier system of *blood* ties. In the other, with blood-relationship perceived, we find the recognition of it stifled, and the Father's Power in the place of blood as the actual measure of relationship. If the former ignores one half of the actual blood ties, viz., those arising through males, the latter as absolutely ignores the other half, viz., those arising through females; and it acknowledges no blood tie, even through males, merely because it is a blood tie. In the one, the system of relationship is natural though imperfect; and that of the other seems almost unnatural—that is, to take the best view of it, it appears to be highly artificial.

That there is evidence of the prevalence of a family system earlier than the Patriarchal Family—evidently and almost necessarily earlier if it generally existed at all—the incidents involved in which explain the actual constitution of societies, which the Patriarchal Theory cannot do, is one reason more for carefully scrutinising the evidence for every proposition which is with that theory associated.

This is not the place to speak at length of the family founded on kinship through women only, or to set forth the evidence of its prevalence. It is not introduced here with the purpose of overbearing the Patriarchal Theory with counter-evidence. The evidence for it, indeed, has never yet been fully exhibited. But much of that evidence is before the world. The portions of the earth "discovered," as we say, within the last three hundred years, supply it in abundance. The study of the ancient nations, too—those very nations upon observation of which the Patriarchal Theory has been founded—has yielded an amount of it which is by no means inconsiderable, and which inquiry is steadily augmenting.

Sir Henry Maine has stated in one of his later books that, even if there was an earlier kinship than the kinship based upon Power, the fact could affect

Comparative Jurisprudence only remotely *—so that, so far as it is concerned, the inquiry whether there was an earlier kinship need not be followed up. But if Comparative Jurisprudence, in tracing the history of society, neglects the earlier history of society, and begins where it is convenient to begin, where a beginning can be made without much trouble, or at some arbitrarily chosen starting-point, does it not expose itself to very serious chances of mistake? Surely it must be most unsafe to assume that nothing anterior to the period of law-books, nothing that does not bulk largely in early law-books, can have had any share in the forming of societies. The conclusions of a science that permits of any such assumptions, must be, it would seem, very liable to error, and always liable to be upset by the results of more thorough inquiry.

All that is now suggested, however, is that the proof from early law-books and other such sources for the propositions involved in the Patriarchal Theory ought to be full, clear, and strong.

* *Early History of Institutions*, p. 57.

CHAPTER IV.

PLAN OF THIS WORK.

Examining the evidence propounded in support of the Patriarchal Theory, we find it to consist in some measure of direct proof of the existence of the Patriarchal Family as Sir Henry Maine has described it, and in a much greater degree of evidence of certain facts from which the former existence of that institution is held to be a legitimate if not a necessary inference.

As to the direct evidence, Sir Henry Maine thinks he has found the Patriarchal Family of the Roman type, with Patria Potestas for its leading feature, (1) among the Hebrews, (2) among the Hindoos, (3) among the Slavonians, (4) among the Irish. It is proposed in separate chapters to examine the evidence adduced, or statements made, with respect to each of these peoples; and on the other hand—such evidence as Comparative Jurisprudence allows appeal to being alone insisted upon—to see what is really indicated by the evidence we possess about each.

Passing to what may be called the indirect evidence, the first and by far the most important branch of it consists of the prevalence of Agnation. Premising that the Patria Potestas in its normal shape has not been, and could not have been, a durable institution, Sir Henry Maine argues that agnatic relationship implies its former existence; and states that, whereas cases of Patria Potestas are rare, agnatic relationship is discoverable almost everywhere, and that, in fact, it would be difficult to say where it has not existed. The inference from this, of course, is that Patria Potestas has existed almost everywhere, and that, virtually, it may be taken to have existed everywhere.

It is proposed to consider at some length the connection alleged to exist between Agnation and Patria Potestas; and, as a preliminary thereto, to discuss—though necessarily in a brief and fragmentary way—the conditions upon which, in various circumstances, the connection between what we take to be a derivative institution and that from which we think it is derived can be held to be established. A full and satisfactory discussion of this subject would be of use in many inquiries. That which will here be ventured upon will, it is hoped, be found not to be without bearings upon the Patriarchal Theory. It may perhaps suffice to show that certain propositions made in *Primitive Marriage* with which the Patriarchal Theory has to reckon—that, for example, which connects the Form of Capture and actual capture, and that which connects the Levirate with Tibetan polyandry—stand, as regards the conditions of proof, upon a very different footing from the connection alleged to exist between Patria Potestas and Agnation. A different account of the origin of Agnation will then be put forward, and supported as far as can be done in a work the purpose of which is chiefly critical. For the introduction of this theory of Agnation apology can scarcely be necessary. Its introduction will be fully justified, it would seem, if it can be thought at all fit to compare with Sir Henry Maine's account of Agnation. And, in so far as it goes to show that Agnation was not a primary form of kinship, it is directly subversive of the Patriarchal Theory.

After this, the cases of Agnation which have from time to time been specified by Sir Henry Maine will come up for examination. It will appear that there is not one of them, the Roman case excepted, in which relationship was clearly agnatic. And, indeed, there is no proof that, before the date of the Twelve Tables, Agnation was established even among the Romans outside the Patrician class. Certain admissions incidentally made in Sir Henry Maine's latest work, *Early Law and Custom*, may, if we choose, save us trouble at this point; and, indeed, it will be found that the chapters on Patria Potestas unavoidably treat of Agnation also for some of the peoples with which they are concerned. The work just mentioned, however, puts at this point an additional labour upon us. In it—in the case of the Hindoos—Sir Henry Maine has shifted his ground, not founding upon the usual data of Comparative Jurisprudence, the law-books, or even upon the earlier Hindoo writings, but taking his departure from

Ancestor-Worship. It becomes necessary to examine minutely the account which he has thereupon given of the Levirate and of certain curious varieties of sonship—curious, that is, as occurring among a people said to have been "Patriarchal"—acknowledged among the Hindoos. And, that done, an attempt will be made to show what the basis of family right among the Hindoos really was. If it is successfully made out that Sir Henry Maine has failed to account for the Levirate and Hindoo sonship in general—the Patriarchal Theory having to reconcile itself with these—the introduction of the former part of this discussion (but that, indeed, could not be avoided) will forward the purpose of this work. And, in so far as, by the latter part of it, it is made probable that the system of kinship through females only can be descried among the prehistoric Hindoos, this part of it is subversive of the Patriarchal Theory.

The other facts from which Sir Henry Maine has made inference of Patria Potestas are the tutelage of women and the heirship of slaves. But it will appear that these need not long detain us.

CHAPTER V.

PATRIA POTESTAS AMONG THE HEBREWS.

THE passage in *Ancient Law** which affirms that Patria Potestas existed among the Hebrews is as follows:

"The effect of the evidence derived from Comparative Jurisprudence is to establish that view of the primeval condition of the human race which is known as the Patriarchal Theory. There is no doubt, of course, that this theory was originally based on the Scriptural history of the Hebrew Patriarchs in Lower Asia. . . . . The chief lineaments of such a society [that is a society organised on the patriarchal model], as collected from the early chapters of Genesis, I need not attempt to depict with any minuteness, because they are familiar to us from our earliest childhood. . . . . The points which lie on the surface of the history are these: The eldest male parent—the eldest ascendant—is absolutely supreme in his household. His dominion extends to life and death, and is as unqualified over his children and their houses as over his slaves; indeed, the relations

* Pp. 122–124.

of sonship and serfdom appear to differ in little beyond the higher capacity which the child in blood possesses of becoming one day the head of a family himself."

Further details are added—of course on the authority of Scripture, but without any particular Scripture being referred to—as to the father owning the flocks and herds of all his children, and as to the law of inheritance. In the sentences omitted from the above citation there is a reference to the controversy between Locke and Filmer, and a suggestion that there is no race of men that was not originally organised on the patriarchal model.

Sir Robert Filmer preceded Sir Henry Maine in alleging, on the authority of Scripture, that Patria Potestas existed among the Hebrews, and he set forth the Scriptures on which he relied in support of his contention, which the later writer thought it unnecessary to do. To those who have studied the controversy between Locke and Filmer* it may seem wonderful that the truth of Filmer's main position could be thus lightly

* The reader who wishes to read this controversy for himself will find Locke's part of it in *Two Treatises on Government. In the former the false principles and foundation of Sir Robert Filmer and his followers are detected and overthrown. The latter is an essay Concerning the true original extent and end of Civil Government.* No author or publisher is named. The book was printed in London, in 1690. In the second essay will be recognised at once Locke's famous Essay on Government. Filmer's *Patriarcha, or the Natural Power of Kings* [*by the learned Sir Robert Filmer, Baronet*] was printed in London, in 1680. It is bound, with separate title and paging, in one volume [which is in the London Library] with Filmer's *Observations Concerning the Original and various Forms of Government.* London, 1696.

assumed by any one, and especially by any lawyer, who had read Locke's masterly reply to the pleadings of his opponent.

The question, however, being whether the Scriptures do prove Patria Potestas or not, a short notice of the matters discussed between Locke and Filmer will carry us some way towards the settlement of it.

The main facts relied on by Filmer for Patria Potestas in Israel were: (1), The sentence passed by Judah on Thamar; (2), Abraham's league with Abimelech; (3), Abraham's army of three hundred and eighteen soldiers of his own family; (4), The fifth commandment, which Filmer takes to be the law enjoining obedience to kings, and of which he says: It is delivered in the terms, "Honour thy father," as if all power were originally in the father. He inferred that the father was absolutely supreme in his household, that he had the power of life and death over his children, the power of making war and peace, and generally all the powers of a King.

Locke, replying, pointed out, (1) as to the injunction on which the duty of obedience to kings was founded, that it was got by only partially quoting the commandment, which is "Honour thy father and thy mother;" and that, by the same method, it could be as easily and as conclusively shown that all power was originally in the mother. As illustrating the position of mothers among the Hebrews, he cited Exodus xx. 12, "Honour thy father and thy mother;" Exodus xxi. 15, "He that smiteth his father or his mother shall surely be put to death;" Leviticus xx. 9, "Every one that curseth his father or his mother shall surely be put to death;" Leviticus xix. 3, "Ye shall fear every one his mother and his father"—in which the woman is named first; Deuteronomy xxi. 18–21, setting forth how a son guilty of habitual disobedience to the voice of his father or the voice of his mother was to be dealt with; and other texts to the same effect.

(2) As to the case of Judah and Thamar, Locke distinguished between an act and the right to do it. "Absalom," he said, "pronounced sentence against Amnon and executed it too." He might have said that what Judah did was no more than to declare the well-known punishment of the offence. There was a *trial*; and the woman was acquitted. More to the point is the observation that Judah was a younger son, whose father and elder brethren were alive. He was not the Paterfamilias. It was not he who, according to the Patriarchal Theory, should have had the power of life and death. If he had this power, how, consistently with the theory, did he happen to have it? "Any man," says Locke, "as well as Judah might have right of dominion."

(3) In Genesis xlii. and xliii. is the story of Israel's trouble about Benjamin, which shows the whole family still clustered round the father. "Reuben," Locke remarked, "offered his two sons as pledges, and Judah was at last surety, for Benjamin's safe return out of Egypt—which all had been vain, superfluous, and but a sort of mockery if Jacob had had the same power over every one of his family as he had over his ox or his ass,

as an owner over his substance, and the offers that Reuben and Judah made had been such a security for the returning of Benjamin as if a man should take two lambs out of his lord's flock and offer one as security that he will safely return the other."

Putting aside the league with Abimelech, and the magnitude of Abraham's following, as, by themselves at least, illustrating nothing except the might and the independent position of the patriarch, it will be seen that Filmer's evidence for Patria Potestas among the Hebrews consisted of the fifth commandment, mutilated so as to distort its meaning; and of what he (wrongly) took to be an example of the exercise of the power of life and death, not by a Patriarch, but by a person who, had Patria Potestas really existed, would have been subject to Patria Potestas himself. Filmer, that is, with all the will in the world to find evidence for Patria Potestas among the Hebrews, in fact found no evidence of it whatever. It need scarcely be pointed out that the fifth commandment throws no light upon the powers possessed by the Hebrew father over his family (though had he possessed the Roman Patria Potestas the commandment might have been, as regarded him, unnecessary); nor that it does show, as many other facts do, that a very high position was assigned to the mother—a position very different from that of sister to her children, assigned to her by the Patriarchal Theory.

As to the story of Benjamin, upon which Locke commented so acutely, it, at any rate, does not read as if the father had the power of life and death over his grown-up sons. The position of the sons seems very distinguishable from serfdom. They show much deference for their father, no doubt, but they address him like men who have a right to be listened to, and, for the general good, press him, and almost coerce him, into a course he was most averse to. And, if any inference is to be made from the story, it can only be that suggested by Locke—that Reuben's sons were his own and not his father's, and that Jacob had nothing like the powers of an owner over Judah—in other words, that Jacob had not Patria Potestas.*

Further on in the history of the Israelites there is no doubt whatever that the father had not Patria Potestas. In Deuteronomy xxi. 18–21, provision is made for the case of a man having a stubborn and rebellious son, who will not hear the voice of his father or the voice of his mother, and whom they have in vain endeavoured by chastisement to correct. The father and mother are to lay hold on the son, and bring him before the elders of his city and unto the gate of his place, and charge him before the elders with his offence. The elders being satisfied, "all the men of his city" were to "stone him with stones that he die." It will be noted that father and mother were both required to come forward, and that they appeared as accusers merely. The idea of Patria Potestas is here excluded. The mother was as necessary to the proceeding as the

* Giving a son as a hostage is, of course, no proof that the father has the admitted power of life and death over his son. But a power extending to life and death over *young* children only would not be Patria Potestas.

father, which by itself excludes Patria Potestas; and the power of life and death was with neither of them, nor with both together, but the customary sentence was executed by all the men of the city after a hearing before the elders.

Going back beyond the time of the Patriarchs, we find a piece of evidence which seems absolutely contradictory of the whole Patriarchal Theory. The first reference to marriage in the Scriptures (Genesis ii. 24) mentions father and mother in a breath, and involves that their son left them when he married. "Therefore shall a man leave his father and his mother and shall cleave unto his wife." The words, ascribed to Adam, must be taken as embodying very early custom. Now what do they mean? Can leaving father and mother (not the father only, observe) mean less than leaving the household, leaving the family, of one's birth?

On the Patriarchal Theory, however, a man, when he married, did not leave the household of his birth. He was not separated from his father and mother; he continued in their family, subject to his father's power, in a condition scarcely distinguishable from serfdom. It was the woman whom he married who left father and mother—giving up all relationship with them—to cleave unto her husband, to become a member of his father's family, to become, with him, subject to his father's Power. In a bride's case, on the Patriarchal Theory, there was a real leaving. On her husband's part there was none.

Even if the passage mean only—what, no doubt, it has commonly been taken to mean—that a son, on marrying, became emancipated, or, in familiar phrase, set up house for himself, it is contradictory of the Patriarchal Theory. For in that case it involves that after getting a wife, at any rate, a man was free from his father's authority; while Patria Potestas extended to all descendants and lasted as long as the oldest ascendant lived.

But, indeed, unless by a man leaving his father and his mother and cleaving unto his wife it can be meant that he continued in his father's family (his mother ranking in it as his sister) and brought his wife to live in it, there is complete discordance between the Patriarchal Theory and the earliest Scriptural record of marriage custom.

This is all we are here concerned with. But it will be well to go a step further and point to a marriage system, altogether different from that of the Patriarchal Theory, which may be that which is indicated in the passage in Genesis. This marriage system seems to be, in a subsequent portion of the Book of Genesis, clearly disclosed as occurring among kindred of the Israelites; and, in the fullest sense of the words, it makes a man leave his father and his mother to cleave unto his wife.

It is what is known in Ceylon as beenah marriage —by which name, as having already been applied to it, let us call it. In beenah marriage the young husband leaves the family of his birth and passes into the family of his wife, and to that he belongs as long as the marriage subsists. The children born to him belong, not to him, but to the family of their mother. Living with, he works for, the family of his wife; and he commonly gains his footing in it by service. His marriage involves usually a change of village; nearly always (where the tribal system is in force) a change of tribe—so that, as used to happen in New Zealand, he may be bound even to take part in war against those of his father's house; but always a change of family. The man leaves father and mother as completely as, with the Patriarchal Family prevailing, a bride would do; and he leaves them to live with his wife and her family. That this accords with the passage in Genesis will not be disputed.

It is in Africa that beenah marriage is now most prevalent; and there are parts of Africa in which it is quite commonly met with—usually alongside of, and, in some sense, contending with, a system of marriage by purchase of the bride and her issue—the two systems, indeed, being generally in use even among the same people, the one preferred in some cases, the other in others.

Its occurrence must be familiar enough to students of works of travel, and it would be superfluous here to accumulate examples of it.* What is more to the purpose is to point out that Jacob made a beenah

* Reference may be made, however, to Marsden's *History of Sumatra*, because of the very interesting and instructive account it gives of beenah marriage as practised in that island.

marriage into the family of Laban; and that Genesis xxiv. 1–8 shows that it was thought not improbable that Isaac, as a condition of marrying into his father's kindred, might have to do the same.

Keeping to the former case, as being the clearer and of itself sufficient, we find, first, that Jacob had to buy his place in Laban's family as husband of Laban's daughters, by service; and, second, that the children born to him belonged to Laban's family, and not to him—both notes of beenah marriage, and the second denoting it beyond possibility of mistake. Jacob, his wives concurring, *stole away* with them and their children from his father-in-law. And Laban, when he had overtaken him, claimed both the wives and their children as his own. "These daughters," we find him saying (Genesis xxxi. 43), "are my daughters, and these children are my children." And further on, after he had agreed to let them go, it is said that "Laban rose up and kissed his sons and daughters, and blessed them." It is easy to understand how they were Laban's and not Jacob's. What could have made them Jacob's was purchase; and Jacob had not purchased. In Laban's days no marriage arrangement was at all likely to be made that was not well known and sanctioned by custom; and, therefore, it must be taken that beenah marriage was—to what extent cannot be known—customary in the land of Haran. The case of Rebekah proves that it was not exclusively practised—for she was purchased, and left the family of her birth. But, as has been mentioned, beenah marriage is now seldom found without marriage by purchase of the bride being found alongside of it—in some places the one, in other places the other, being the more in vogue; while, quite commonly, the one or the other, according to circumstances, is preferred in different cases by the same people.

If it be beenah marriage that is indicated in Genesis ii. 24 there is at once an end of the Patriarchal Family, so far as the Hebrews are concerned, and therefore an end of it as a universal and primordial institution. And if Jacob's was a beenah marriage—which can scarcely be seriously disputed—why doubt that beenah marriage is there indicated? Marriage by purchase of the bride and her issue can hardly be thought to have been primeval practice. When we find beenah marriage and marriage by purchase as alternatives, therefore, it is not difficult to believe that the former is the older of the two, and that it was once in sole possession of the field.*

Putting aside this question of beenah marriage, however, the story of Laban and his family at any rate enables us to decide whether the Patriarchal Family, with its incidents of Patria Potestas and Agnation, existed among the early Hebrews.

On the Patriarchal Theory, Laban's family should have consisted only of his sons and their descendants and his unmarried daughters. His daughters should have been cut off from him by marriage; and their children, because included with them in another man's

* As to this, see p. 273, *et seq.*

family, should have been as strangers to him.* Laban's daughters, their husband, and their children, all lived with him, however, and he continued to regard his daughters as daughters, and counted their children among those belonging to him. "The organisation of primitive societies would have been confounded," says Sir Henry Maine, if men had acknowledged relationship through women. Here is Laban, nevertheless, claiming his daughter's children as his own. And, before that, he had at once acknowledged relationship between himself and his sister Rebekah's son. "Surely," he had said to Jacob, "thou art my bone and my flesh." Need it be said that Laban's was clearly not a "Patriarchal" family? Laban had never dreamed of Agnation. What authority he claimed over his household does not appear; but whatever it was, the limits of its application were quite different from those of Patria Potestas.

* Jacob's wives, justifying to themselves the proposed desertion of their father's household, say, "Are we not counted of him (their father) as strangers? For he hath sold us." But clearly this is only said to justify to themselves what they were going to do. In fact, they had all along been members of Laban's household; and the sequel shows that they were not counted by him as strangers. Plainly he had not *sold* them. If he had done that, Jacob's right to go away with them and their children would have been unquestionable, and it is clear that he knew it to be open to question. Their initial outburst, "Is there yet any portion or inheritance for us in our father's house?" discloses a state of mind impossible to women brought up in a "patriarchal" family. Such women would take it as of course that, once married, they could have nothing to look for from the family of their birth. Laban's daughters speak, be it observed, of the property Jacob was about to carry off as *theirs* and their children's, seeming so to justify his taking it away.

Jacob, on his part, had gone to his mother's brother, feeling sure he would find protection as a relative. He was most kindly received. In his marriages he only had to submit to the custom of the country. In his circumstances, there could be no question of his getting a wife, as his father Isaac had done, by purchase. And, as bearing on Patria Potestas, it should be noticed that, by the custom of the country, a family which took in a man to be husband to one of its daughters, instead of compensating the man's father for the loss of Potestas over him, exacted a price in service for his admission. Patria Potestas must, therefore, have been unknown in Haran. One can the more easily believe it was unknown in the household of the husband of Laban's daughters. We have seen that, in Jacob's case, there is literally not a trace of it.

Of minor facts which go to show that Patria Potestas was unknown among the early Hebrews and their kindred, only one or two need be added.

And, first, the concentration of all family property in the hands of the Paterfamilias being among the features of Patria Potestas, what is to be made of the fact—again from Haran—that at the espousal of Rebekah, the bridal gifts (the bride's price) were given, not to the father of the bride, but to her brother and her mother? The father, if he had had Patria Potestas, ought to have got them all, as "compensation for the Patriarchal or Family authority which was transferred to the husband;" and, indeed, his wife and son should have had no property independently of him. But he got none. The wife and son *were* capable of having property, and it was they who were compensated. Not to dwell upon the other bearings of this—which, indeed, are obvious enough—nothing could be more conclusive against Patria Potestas.

Again, we find that each of Laban's daughters, on her marriage, got a maid from her father; and the maid in each case seems to have become the daughter's property, of which she could dispose at her own will. Each daughter gave her maid to Jacob to bear children for her, and the children were counted with those of the mistress.* Similarly, in the house of Abraham, Sarah gave to her husband her Egyptian maid to wife; and though Abraham, had he been a true Paterfamilias, should have been king in his own house, it would seem as if, without Sarah's consent, he could not have begotten Ishmael. Afterwards, on his wife's order, he had, sorely against his will, to turn Hagar and Ishmael out of doors. It does not look as if Sarah was *in manu*.

As bearing on the law of succession, and the representative character sustained in primitive times, on the Patriarchal Theory, by the head of the family, observe that Abraham, though he had sons after Sarah's death, left all that he had to Isaac, dismissing his other sons with gifts in his lifetime. Abraham seems to have acted as full proprietor of his estate.

As to Agnation, besides the evidence of its non-existence already adduced, there is abundant Scriptural evidence to show that, instead of its being established, the

* As to this, see note on page 273.

relatives on the mother's side were anciently the closest kindred. One case will prove this as well as twenty, and, to take the first that comes to hand, see Judges ix. 1–4. Of course Abimelech's pretension to rule arose out of his being a son of Gideon. But it was his mother's family that helped him to power, and they did so because he was "their bone and their flesh." It is Patria Potestas that is here being dealt with. But what is the Patriarchal Family without Agnation?

Sir Henry Maine says in one place* that the connection of the Patriarchal Theory with Scripture was rather against its reception as a complete theory, because most of the early inquirers into social phenomena were either under the strongest prejudice against Hebrew antiquities, or were strongly desirous of constructing systems without the assistance of religious records. The Scriptures make it impossible, however, to accept as a complete theory the Patriarchal Theory as it has been enunciated by himself. Plainly, the Hebrews must be excepted from it. The Scriptures not only do not countenance it, but they contradict it, so far as the Hebrews are concerned. It is not merely that they contain nothing to suggest that the Family founded upon Power ever existed among the Israelites. All the evidence there is goes to show that the Roman institution of Patria Potestas never prevailed among them; and, as regards their early history, which is what concerns us, we have incidentally found complete disproof of Agnation. It is needless to dwell upon

* *Ancient Law*, p. 122.

the indications that have appeared of their having had in very early times a family system almost the converse of the Patriarchal—though these must count for something in estimating the claims of the Patriarchal Theory to acceptance. So much may be said, notwithstanding that it has to be borne in mind that Sir Henry Maine has not adduced the evidence from Hebrew antiquities and religious records upon which his opinion appears to have been formed. As to impressions remaining with us "from our earliest childhood," too much must not be made of them. It is certain, however, that, in spite of such impressions, the picture of the Hebrew family sketched in *Ancient Law* comes upon most people as a surprise.

CHAPTER VI.

PATRIA POTESTAS AMONG THE HINDOOS.

In *Ancient Law* Sir Henry Maine has stated that the greater part of the legal testimony to the truth of the Patriarchal Theory comes from the institutions of the Romans, Hindoos, and Slavonians; adding—as if under pressure of the evidence at his disposal—that "the difficulty at the present stage of the inquiry is to know where to stop, to say of what races of men it is *not* allowable to lay down that the society in which they were united was originally organised on the patriarchal model." His description of the Patriarchal Family immediately follows—the eldest male parent, the eldest ascendant, absolutely supreme in his household, his dominion extending to life and death, and being as unqualified over his children and their houses as over his slaves, while there is no property in the hands of any of them that is not really his. It is quite clear, then, that he considered Patria Potestas to have been a feature of the family system of the Hindoos, though he does not in *Ancient Law* affirm that there is direct legal testimony to that effect. In his *Early*

History of Institutions (p. 323), however, he says that "The Hindoos may be as confidently asserted as the Romans to have had their society organised as a collection of patriarchally governed Families;" and, in another place (p. 310), that the two societies of Rome and India "are *seen* to be formed at what, for practical purposes, is the earliest stage of their history, by the multiplication of a particular unit or group, the Patriarchal Family." And a description of this unit or group follows the latter passage: "The group consists of animate or inanimate property, of wife, children, slaves, land and goods, all held together by subjection to the despotic authority of the eldest male of the eldest ascending line, the father, grandfather, or even more remote ancestor." In the latter passage he does seem to speak as if there was direct evidence of Patria Potestas among the Hindoos. At any rate, the prevalence of that institution among them has always been "confidently asserted" by him, and always spoken of as if there was evidence for it of some sort that must compel belief.

Let us see, then, what Hindoo law-books—since it is to law-books that Comparative Jurisprudence makes appeal—show as to the existence or non-existence of Patria Potestas among the Hindoos.

This is not the place to discuss the comparative antiquity and authority of Hindoo law-books. It must suffice to say that, until recently, nobody ever doubted that the most ancient and the fullest statement we have of early Hindoo law is contained in the Manava

Dharma Shastra, best known to us as the Code of Manu; and that, though this is now in question, the reasons for allowing the work the authority that has been assigned to it do not as yet appear to have been shaken. From it will now be cited some passages which bear upon Patria Potestas:

1. "Let no father who knows the law receive a gratuity, however small, for giving his daughter in marriage, since the man who . . . takes such a gratuity is a seller of his offspring."—*Manu*, chap. iii., ver. 51.

From this it appears that a Hindoo father had no power to sell his offspring.

2. Chap. iv. (On Economics and Private Morals), ver. 180, declares it to be the duty of a housekeeper to have no strife with his mother, father, son, wife, or daughter. In vers. 184 and 185 of the same chapter, it is declared that he must consider "his elder brother as equal to his father, his wife and son as his own body, his assemblage of servants as his own shadow, and his daughter as the highest object of tenderness." When offended with any of them, he is told to bear the offence without indignation.

It was the duty of a housekeeper, that is, to be self-restrained, forbearing, indulgent towards his family.

3. From chap. viii. (On Judicature) it appears that all jurisdiction was vested in the king or his judges, and in the father of a family none whatever. "Altercation between man and wife, and their several duties" is named (ver. 7) as one of the eighteen principal titles of

law to be daily dealt with by the courts. That is, altercations between husband and wife, and questions respecting their several duties were matters for settlement in the courts of justice.

Here it may be seen that the Hindoo father was not "absolutely supreme in his household," and that his wife was not *in manu*. If he had a dispute with his wife, the courts decided between them.

And the jurisdiction of the courts extended to all the members of the family alike, and to all offences which they could respectively commit against one another. "Neither a father, nor a preceptor, nor a friend, nor a mother, nor a wife, nor a son, nor a domestic priest must be left unpunished by the king if they adhere not with firmness to their duty."—(*Manu*, chap. viii., ver. 335.)

It appears from the same chapter that a father had a certain power of correcting the members of his family. "A wife, a son, a servant, a pupil, and a *younger* whole brother may be corrected, when they commit faults, with a rope or the small shoot of a cane; but on the back part only of their bodies, and not on a noble part by any means. Who strikes them otherwise than by this rule incurs the guilt *or shall pay the fine* of a thief." (Chap. viii., vers. 299, 300.) On the former of these verses Mr. Colebrooke observes:* "May I quote a maxim of no less authority? 'Strike not, even with a blossom, a wife guilty of a hundred faults.'" But, at any rate, these verses show that the

* *Hindu Digest*, Vol. II., p. 209.

Hindoo father's power over his family did not extend to life and death. Even the power of correction allowed him—a power which, as regards children, fathers probably have among every people—was carefully limited, and a penalty prescribed for any abuse of it. From the same chapter (ver. 335, already quoted, and ver. 389) it appears that, while his powers over his family were thus limited, he was under strict legal obligation to his family to stand by and support it, and that he was liable to punishment if he did not, unless he could plead, in regard to members of it whom he forsook, that they had been guilty of deadly sin.

It may now be said that, as regards its essential elements, the Code of Manu shows the Hindoo father not to have possessed Patria Potestas, or any power approaching to it, or capable of suggesting it. If the prescriptions of "the law" were of any avail, he was without the powers of life and death and sale. It remains to see whether all the property of the family belonged to the father, or whether its members could have property independently of him. And—

1. Manu ix. 194* says: "What was given before the nuptial fire, what was given on the bridal procession, what was given in token of love, and what was received from a brother, a mother, or a father, are considered as the sixfold *separate* property of a *married* woman." Property given to a woman

* Sir William Jones's translation. The words in italics in this and other extracts are from the gloss of Kulluka, which Sir William followed.

on her marriage was "inherited by her *unmarried* daughter" (Manu ix. 131). As to her other property, on her death, "the uterine brothers and the uterine sisters, *if unmarried*," were to "equally divide the maternal estate" (ix. 192). A further provision is (ix. 195) that "What she received after marriage from the family of her husband and what her affectionate lord may have given her"—that even that—was to be "inherited, even in his (her husband's) lifetime, by her children." On the other hand, "Of a son dying childless, *and leaving no widow, the father and* mother" were "to take the estate" (ix. 217). When the wife herself died childless, in certain cases the husband inherited her property, while in others—so far was a wife from being cut off by marriage from the family of her birth—it went to her father and mother (ix. 196, 197). The Code enjoins on a woman not to make hoards from the goods of her kindred, or even from the property of her lord without his assent. It further reserves to her what we call paraphernalia.

The sixfold enumeration of a married woman's separate property appears not to have been restrictive, but the precise limits of that property at the time when the text of Manu was settled do not now concern us. It is enough to know that it was fully recognised, and that, from the notice it receives, it must often have formed an estate of considerable importance. It, at any rate, included the *dos* and the nuptial gifts of the husband and his family, and all property given to the wife after marriage by her own family, or by her husband and his family. In all such property of the wife the husband had no right whatever. At the wife's death it went to her children, even if the husband was then living; and even if she left no children it did not in every case become his. When it went to her sons and daughters it was their separate property just as it had been hers; which shows that sons and daughters, as well as wives, were capable of holding property, and must often have had property, independently of the head of the family. For further proof of this, take the provision made for the son of a Sudra woman by a man of one of the three higher classes (ix. 155). Such a son was to inherit no share of the family estate, but whatever the father might give him was to be his. He might be provided for, that is, by a gift *inter vivos;* and, therefore, he was capable of holding property in his father's lifetime. Still more must sons belonging to the higher classes have had that capacity.

It now appears that the whole property of the family did not belong to the father, and therefore that there was no element of Patria Potestas to be found among the Hindoos at the date of Manu. But further, as to property:

2. Whether the sons of a family were co-owners with their father in the undivided ancestral estate, or had only a right of sharing in it at his death, or when he divided it, has been much contested among Hindoo jurists. There is evidence (but not in Manu) that they could enforce a division of it against his wish. Gautama—alleged by Mr. Bühler to be an older writer

than Manu, and no doubt a great authority—enumerates sons who had done this among the classes of people who were not to be allowed to partake of funeral oblations;* on which Mr. Bühler remarks: "From this sutra it would appear that sons could enforce a division of the ancestral estate against his [their father's] will, as Yagnavalkya also allows (see Colebrooke, Mitakshara, I. 6, 5–11), and that this practice, though legal, was held to be *contra bonos mores.*" A power of enforcing partition would, it should seem, prove the joint ownership of sons. It would conclusively prove the absence of Patria Potestas. What Manu lays down as to the proper period of partition is as follows: "After the death of the father and the mother, the brothers may divide among themselves the paternal and *maternal* estate; but they have no power over it while their parents live, *unless the father choose to distribute it*" (Manu ix. 104). Here the period of partition is postponed to the death of the mother in the event of her surviving the father—which also is a provision inconsistent with Patria Potestas. For if the verse proves that sons had no power over the estate so long as the father lived, it proves equally that they had no power over it, after his death, while their mother lived.† But

* *The Sacred Laws of the Aryas.* Translated by Georg Bühler, Oxford, 1879. Gautama's *Institutes of Sacred Law,* ch. xv., s. 19.

† For authorities, and an argument founded on them, to show that after the father's death the mother was anciently head of the family, possessed, with full right of control, of the family estate, see Mr. J. D. Mayne's *Hindu Law and Usage,* Madras and London, 1878, pp. 124, 125.

it may merely lay down "a precept of perfection," and be not inconsistent with the sons having the power to enforce a partition if they were so wicked as to wish to do so. It can scarcely be thought that property rights in the estate were not possessed by them after their father's death. But if they were then possessed by them, there is nothing to show that they were not possessed by them in his lifetime. It is to be gathered from the older Hindoo text writers that partition was usually made with the father's consent in his lifetime, and that it was thought proper he should make it. He divided not only property he had inherited, but also his own acquisitions; keeping a share to himself, and being allowed some choice in the distribution of his acquisitions.

3. About the earnings of the members of an undivided family after the death of their father, there is a good deal to be found in the Code of Manu; and we learn that each brother could keep to himself, if he pleased, wealth acquired by him without using the patrimony; and that brothers who put their earnings into the common stock (whose gains may from their so doing be supposed to have been nearly equal) were to have the property acquired divided equally between them—to the exclusion of the slight preference which might be claimed by the eldest in a division of the patrimony (ix. 204–9). The undivided brothers, in short, were free to choose between holding each by his own earnings, and making hotch-pot with one another by throwing their earnings into the patrimony —the eldest, though having the control, getting no

advantage in a division over the others in the latter case. As to the earnings of an undivided family while the father lived, we find a rule laid down for one case only (which may have been the commonest). It is that "if among undivided brethren *living* with their father, there be a common exertion for common gain, the father shall never make an unequal division among them *when they divide their families*" (ix. 215). This reads as if the augmentation of the patrimony by gains arising from the common employment not only involved that such gains should be equally divided, but excluded the father from showing any preference in the distribution of the patrimony itself. But taking it to imply that there was to be an equal division only of the family gains, it would be identical with the provision made for the case of brothers who held together after their father's death and made a common stock of their gains; and this shows that, before his father's death—just as after it—a man had assured to him his fair interest in the surplus profits of his labour which accumulated in the hands of the head of the family. What happened when, the father being alive and the family undivided, there was not a common employment, Manu does not tell us; and it does not seem worth while to go to more modern authors for light upon the matter. It may be inferred that the father was not bound to make an equal division of acquisitions in this case; but that, if the earnings of his sons came into his hands, he would be expected to make a just division; and that is all that concerns us.

As regards accumulations from joint earnings which were in the father's hands, then, as well as in regard to the ancestral estate, there was between father and sons a sort of joint ownership; while a son was capable of holding, and must frequently have held, other property as separate property. That is what the Code of Manu shows us. Having shown this, we have shown the absence of every element of Patria Potestas.

It remains, however, to consider a passage (Manu viii. 416, 417)—the only passage in Manu of which so much can be said—which, superficially considered, may seem to carry some implication of one element of Patria Potestas. It is regarded by Hindoo jurists of every school, and beyond all doubt rightly regarded, as bearing upon earnings only. It is as follows:

Ver. 416: "Three persons, a wife, a son, and a slave, are declared by law to have *in general* no wealth exclusively their own; the wealth which they may earn is *regularly* acquired for the man to whom they belong." Ver. 417: "A *Brahmin* may seize without hesitation, *if he be distressed for a subsistence*, the goods of his *Sudra* slave; for, as that slave can have no property, his master may take his goods."

Of course the words in italics, from the gloss of Kulluka, convey what seemed to the commentator necessary qualifications of propositions too wide in their terms. The verses occur in the chapter on Judicature and on Law private and criminal; and, as frequently happens in that chapter, the one is merely a preamble to the other—the first verse prescribing nothing, and being, as is apt to happen with preambles, and as we find to be the case repeatedly in this chapter, a great deal larger than was necessary to cover what the writer desired to lay down. That was, that the owner of a slave need not hesitate to appropriate the slave's peculium, since, strictly speaking, all a slave had was his master's—a proposition harsh-sounding and liable to abuse; so that the commentator had to attach to it qualifications which confine it to the Sudra slave of a Brahmin master, who happened to be himself in distress. It is, however, only with the preamble to this allowance of the extreme right of a master that we have to do, and of course with that only in so far as it refers to the position in respect of property of a wife and of a son. And, in the first branch of the verse, they are said (not to have no wealth, but) to have no wealth exclusively their own—which Kulluka has qualified by inserting the words, in general. In the second branch of it, it is alleged that their earnings are acquired for the head of the family; to which Kulluka has attached the qualifying word, regularly—which may be taken to be a variant of the words, in general (regularly, as a rule, in general).

The verse, therefore, does not deny the capacity of wife and son to possess property, but admits it. And, except as to earnings, it does not suggest that the father was even a sharer in the wealth of wife and son which was not exclusively their own.

It must be borne in mind that much—perhaps

most—of the property of Hindoos was in the position here described.* We have an example of property not being exclusively one's own, in the ancestral estate as possessed by a father before he made partition with his sons; for, even if it be denied that his sons were co-owners with him, it is certain that he could not dispose of it at his pleasure. The interest in their common property of individual brothers, when they continued undivided after their father's death, putting their gains into the common stock, is another example of it. The case of brothers living undivided with their father and pursuing a common industry—the only case involving earnings made in the father's lifetime and before partition for which the Code of Manu makes provision—is a third; and it is one of the cases which the writer of the preamble must have had in view.

The separate property of a wife or a son is not, so far as is disclosed in Manu, an example of wealth that is not exclusively a person's own. Such property is declared to be the property of the wife, or of the son or daughter, without any qualification. If, however, in the interest of the heirs to whom it reverted, it could not be wasted, the heirs were co-owners in it. It was not the exclusive property of wife or son or daughter. And yet the father, as such, had no right to it.

* "Among the Hindoos," says Mr. J. D. Mayne, "absolute, unrestricted ownership, such as enables the owner to do anything he likes with the property, is the exception. The father is restrained by his sons, the brother by his brothers, the woman by her successors. If property is free in the hands of its acquirer, it will resume its fetters in the hands of his heirs."—*Hindu Law and Usage*, p. 175.

But, as noticed already, it is not suggested that *all* the property of wife and son which was not exclusively their own belonged to the father so far as it was not theirs. The claim made for *him* is confined to earnings. And the writer of it had earnings, and savings out of earnings, in mind when making his initial flourish about the wife, the son, and the slave having no property exclusively their own, whether he had anything else in mind or not. It was savings out of earnings he would naturally have been thinking of when seeking to find authority for the pillaging of slaves. And most probably he thought of nothing else.

But had he been thinking dimly of something wider —and, no doubt, it was a case of exigency he was about to provide for—had he had in mind that the property of wife or son was liable for debts incurred in time of exigency for behoof of the family (Manu viii. 166), it is to be observed that this liability carried with it no implication of Patria Potestas. For all the family property was liable to be drawn upon in time of exigency. And it was not the father only who could make it liable. The family administrator, whoever he was, could do so. The ground of liability was family necessity, and not paternal right.

As to the father's right to receive the family earnings, it has been shown that he was under regulation as to the distribution of savings out of such earnings in one case—the case, no doubt, of most common occurrence among the Hindoos. It is likely he was not unrestricted as to this in any case. And his sons were, in fact, co-owners with him of such savings. But, at any rate, the father's right to receive earnings does not affect the fact that wife and son had the capacity of holding other property independently of him.

It is now plain that the passage under notice, though words perhaps have not been weighed in it—and it belongs to a kind of writing in which words seldom are weighed—in which to make out somehow the appearance of a reason for that which is to be done is what is thought of—does not carry the suggestion which might in inconsiderateness be attached to it. When we examine it to see whether it can be regarded as claiming the "Patriarchal" power over property for fathers, it becomes clear that there is nothing to be said for that view of it unless we overlook its words, and distort and add to its meaning. For it makes no claim over all the family property for the father. And it is not really at variance with the prescriptions of the law.

As we find that the first branch of the preamble (so to call it) refers to accumulations out of earnings—whether to anything more or not—instead of this passage making against what is elsewhere laid down in Manu as to the property of wife or son, we get from it additional proof that sons had an interest assured to them in the property derived from their labour when it remained in their father's hands. For in saying that such property was not exclusively theirs, the writer admits that they had a property interest in it. And he does not suggest or allow any encroachment upon this—or upon any property of son or wife—but implicitly declares its inviolability.

From the gloss of Kulluka it is obvious that there were cases in which earnings were not in any sense acquired for the head of the family, and therefore cases in which the property arising from them was exclusively the acquirer's own. Accepting the fact—which nobody will question—it is unnecessary here to inquire what the cases were, especially as the Code of Manu does not help us in the matter. What concerned us here was to show that the passage we have been considering carries, as regards property, no implication of Patria Potestas, and that appears to have been done sufficiently.*

It should here be mentioned that after the separation which followed upon a division of the family property, the members of the Hindoo family remained, as they had been before, each other's nearest relatives with possibility of heirship—so that if, after that, the father made acquisitions and had a son, the separated brothers were free at his death to come in and make hotch-pot with the latter. Separation, that is, had not any of the effects of emancipation.

It has already appeared that the Hindoo wife was by no means *in manu*. A few passages may be quoted to show what her position in the family was:—

"The law, abounding in the purest affection, for the

* The history of the family among the Hindoos is considered at some length in chapters xvi. and xvii., on "Sonship among the Hindoos."

conduct of man and wife" lays down, "Let mutual fidelity continue to death," as the supreme rule for the married pair.—*Manu* ix. 101, 103.

"Married women must be honoured and adorned by their fathers and brethren, by their husbands and by the brethren of their husbands, if they seek abundant prosperity. Where females are honoured there the deities are pleased; but where they are dishonoured all religious acts become fruitless."—*Ibid.*, iii. 55, 56—from which also it appears that marriage by no means cut off a woman from her family; but this can be proved by scores of passages.

"He who truly and faithfully fills both ears with the Véda must be considered as equal to a mother; he must be revered as a father; him the pupil must never grieve. A mere *áchárya*, or *teacher of the gáyatri only*, surpasses ten *upádhyáyas*; a father a hundred such áchâryas; and a mother a thousand *natural* fathers."—*Ibid.*, ii. 144, 145.

"He shall be fined a hundred who defames his mother, his father, his wife, his brother, his son, or his preceptor."—*Ibid.*, viii. 275.

"If an elder brother act as an elder brother ought he is *to be revered* as a mother, as a father."—*Ibid.*, ix. 110.

"The wives of his preceptor, if they be of the same class, must receive equal honour with their venerable husband."—*Ibid.*, ii. 210.

"Let every man constantly do what may please his parents; and on all occasions what may please his preceptor. Due reverence to these three is considered as the highest devotion; and without their approbation he must perform no other duty."—*Ibid.*, ii. 228, 229.

"Him by whom he was invested with the sacrificial thread, him, who explained the Véda or even a part of it, his mother, and his father, natural or spiritual, let him never oppose."—*Ibid.*, iv. 162.

Such passages might easily be multiplied, but these are enough to show that, in the Hindoo family, as among the Hebrews, the mother was equal in honour with the father, if not something more. When the two are mentioned together, the mother is always mentioned first. We have seen that the text in Manu relating to partition ordains that sons should live together undivided till after the death of both father and mother; and there is not wanting authority to show that anciently the mother had some control over the family estate.*

It is now clear that the Code of Manu negatives every element of Patria Potestas as regards the Hindoos for the period at which it was drawn up. And it is not being too venturesome to say that in no ancient collection of Indian laws is there a hint of it as actually existing, or as having at a previous time existed in India.

It will occur to most people, no doubt, that the Code

* As to the position of the mother in the Hindoo family, see also the Vivada Chintamani, p. 225; Apastamba i. 4, 14, v. 6; *ibid.*, ii. 2, 4, v. 13; *ibid.*, ii. 6, 13, v. 1–5; and ii. 50, 51. It seems needless to give further authority as to her right to possess separate estate.

of Manu—and the same may be said of all other Hindoo law-books—is concerned with a social state by no means primitive, in which it is idle to look for the beginnings of things. But this is only what is to be expected when we take comparative jurisprudence upon its own ground. Going further back, however, we are not aware that any trace of Patria Potestas, or of the family system of the Patriarchal Theory, has ever been pointed out in the ancient Sanscrit literature. It is perhaps natural to conjecture that the rights of property possessed by wives and children at the date of Manu were once non-existent, and that they grew up by degrees in previous ages. Even were this so, the fact would no more prove that such rights were reared upon Patria Potestas than the history of our own law as to married women's property could prove that English wives were formerly *in manu.* But if we push inquiry back we meet with facts which tend to show rather that a growing down of the rights of property of women and children occurred among the Hindoos than that they grew up. For there is abundance of evidence that a provision like that of the Code of Gentoo Laws (ch. ii. 8, 14), regulating succession in the case of children born to a woman living in polyandry, was much needed in India in the earliest times. This is not the place to go into that evidence, and it must suffice to say that striking proofs of the prevalence of polyandry among early Hindoos of saintly and of princely stock are furnished by the Mahabharata. It will be found a difficult task to reconcile them with the primordial existence of the Patriarchal Family among the Hindoos.

It is proper to add that were evidence of Patria Potestas of later date than Manu forthcoming among the Hindoos, it would not support Sir Henry Maine's propositions. For that would show, not that Patria Potestas is primordial, but that it is not primordial; not that it is early, but that it is late, appearing, if it appear at all, after society has passed through a long course, and institutions have become complex.

CHAPTER VII.

PATRIA POTESTAS AMONG THE SLAVS.

In *Ancient Law* the Slavonians, with the Romans and Hindoos, are said to be the races which furnish nearly all the legal testimony in support of the Patriarchal Theory. What legal testimony has come from the Slavonians is nowhere stated in that work; but it contains some notices of the Village Communities of Russia; and it quotes, as if adopting it, a statement* of some of "the earliest modern writers on Jurisprudence," that it was only the fiercer and ruder of the conquerors of the empire, and notably the nations of Slavonic origin, which exhibited a Patria Potestas at all resembling that which is described in the Pandects and the Code. In what the resemblance here spoken of consisted is not disclosed, nor is it suggested that there was more than a resemblance.

In a much later production than *Ancient Law*,†

* *Ancient Law*, p. 143.

† "South Slavonians and Rajpoots." *The Nineteenth Century*, December, 1877. Reprinted, with some modifications, as chapter viii. of *Early Law and Custom*, under the title, "East European House-Communities."

Sir Henry Maine has identified the House-Community of the South Slavonians, and also the Joint Undivided Family of the Hindoos, with "the Roman Gens [in *Early Law and Custom*, alternatively with either the Gens or the body of Agnates], the Hellenic γένος, the Celtic Sept, the Teutonic Kin;" and has declared himself satisfied—not from knowledge, but as a matter of inference—that the natural families which now occur among the South Slavonians are "Patriarchal" families, despotically governed by the oldest ascendant.

Taking first the statements of the earlier work, their vagueness, which must in any case have made them difficult to deal with, is, as matters stand, peculiarly perplexing. The author does not in any way suggest to us what is the nature of the legal testimony contributed by the Slavonians which he has had in mind; and he does not tell us what authority exists for a statement about the paternal power among the early Slavonians—nor even who are the writers on jurisprudence who professed to have knowledge of its nature or limits. This is peculiarly perplexing, because, on examining what is recorded of the early Slavonians, what one finds is that there is a truly surprising absence of information about their institutions, and, to all appearance, nothing whatever to warrant the statements of Sir Henry Maine and his authorities.

The Slavonic nations began to be known to the Greeks in the latter half of the sixth century, Slavonic tribes having by that time made their appearance in the countries bordering on the Danube. Onwards from that period pretty frequent notices of them occur in the Byzantine writers. These will be found collected in the voluminous work of Stritter. Frequent as these notices are, however, a short passage in Procopius (A.D. 562), with a few sentences of later authors, contain all the Byzantines have preserved to us concerning the social habits or institutions of the Slavonians. Procopius tells us that the Slavs and the Antes did not obey one man, but lived in a state of democracy; that they followed their own ancient customs; that they worshipped the God of Thunder, regarding him as the only Lord of the Universe, but worshipping besides rivers and nymphs and other divinities; that they lived in poor and widely-scattered huts, and frequently changed their place of habitation; and that both peoples had previously been called Spori—as he thinks, because their settlements were so far apart, and they occupied so great a space. The Emperor Constantine Porphyrogenitus, writing some four hundred years later than Procopius, says that the Croats and Servians had no princes except old men, who were their jupans, and that the same might be said of all Slavish peoples; but he tells us nothing of their jupans* except that they were old. From writers intermediate between these two we hear of the love of liberty which distinguished the Slavs, or rather of their repugnance to put up with any master; also of the devotion of Slavish wives to their husbands, which often led them to kill themselves at the husband's obsequies. And when to

* The name is said to be of Gothic origin.

this is added that they liked to establish themselves on river banks, in woods, and in marshes, all has been told that the Byzantine writers have handed down to us. Taken altogether, it does not afford material for even a conjecture about any Slavonic institution.

The Frankish and Northern annalists are no more helpful to us in this matter than the Byzantine historians; and even the Chronicle of Nestor—the oldest Slav Chronicle extant, its author a monk of Kiev in Russia, who lived in the second half of the eleventh and early years of the twelfth centuries—gives us scarcely a hint as to the domestic economy or social system of the early Slavs. Such as he is, however, Nestor is the only early Slav authority worthy of being seriously looked at. M. De Laveleye has relied upon a passage in the "Libusen Sud," or "Judgment of Libusa"—one of the so-called national poems discovered at Königinhof in Bohemia in 1824—for proof of the antiquity of the House-Community among the South Slavonians. But whatever may be the antiquity of this institution, it is not to be forgotten that those poems are of as doubtful authenticity as the poems of Ossian; that the period of their composition is more uncertain still; and that the princess, whose judgment in the case of two brothers who had quarrelled over the division of their inheritance is taken to show that it was Slavonic custom not to divide inheritances, is unquestionably a fabulous person, a creation of the popular fancy—the popular memory, which has preserved the tradition of her and of many other personages equally mythical, having completely lost hold of actual events and real personages of date not much earlier than that to which she is assigned.*

Nestor fixes the coming of Rurik and his Varagian followers into Russia at about two hundred years before his own day (at A.D. 862), and he gives accounts of the early inhabitants of Russia which apply to times more remote. He begins, indeed, with the Flood, and traces, though very briefly, the descent of the Slavs from Japhet; but this need not impair our faith in descriptions which he cannot be suspected of having invented, and which are consistent with what is otherwise known of barbarous peoples. What can be gathered from him that in any degree bears upon the present inquiry can be quickly told.

The country was thinly peopled, and little better than a wilderness. The small communities which formed its population had at first little to do with one another. In later times they sometimes combined against a common enemy, for example to resist the predatory onsets of the Varagians from beyond the Baltic. They had given up wandering, and had settled down to tillage; and they lived in enclosed villages in happy ignorance of the distinctions of rank. They were known to each other by names derived from the physical character of the districts in which they lived, or from the name of some mountain or river. Thus, the people of Kiev were Polians, inhabitants of the plains, and

* She is assigned to the latter part of the seventh century. The Slavs made their appearance in the Danubian countries about a century and a half before.

their near neighbours were Dreyvians or men of the woods, while the Polovzans again were the people who dwelt by the river Polovz. Their country abounding in wild beasts, they were great hunters, and had a wealth of skins—in which, in somewhat later times, they paid tribute to their conquerors, and which, indeed, for centuries continued to form the basis of the Russian currency. With a general resemblance between all, each community had its own peculiarities of custom and manners; and Nestor—himself a dweller at Kiev—gives a much more favourable account of the Polians than he does of any of their neighbours. The Polians, he says, were quiet and gentle. They showed much respect to their parents and their relations, and to their daughters-in-law, fathers-in-law, and brothers-in-law. They had a form of marriage. The bridegroom did not himself go to fetch his bride, but some one brought her to him in the evening, and the price stipulated to be given for her was sent the next day. Like other heathen, they sacrificed to lakes, and springs, and plants. The Dreyvians, on the other hand, lived in a brutal way, like cattle. They killed one another. They ate all unclean things. They really were without marriage (that is, they did not marry after an agreement, like the Polians); but they carried away maidens by force, and took them to their beds. The Radimitschans, the Viatitschans, and the Severians had like customs—living in the woods like wild beasts, and eating all unclean things. They carried on unchaste conversations before parents and daughters-in-law, paying no respect to them. They had no marriage; but they arranged merry games at which they played, danced, and sang devilish songs, and, at the end, each man carried away a woman who became his wife. Those who could get them had two or three wives. They burnt their dead, gathering the remains and putting them into an urn, which they placed on a pillar at the wayside. The Polovzans, again, smudged themselves with blood; ate carrion, moles, marmots, and all unclean things; married step-mothers and daughters-in-law; and "committed all the other wicked practices of their fathers."

Such is Nestor's brief account of the ancient manners of his countrymen. It appears from it that, except among the Polians, who had marriage by purchase, probably with the form of capture, wives were usually got by actual capture; and that there were in use among some of them friendly arrangements, such as have been found among other peoples, to facilitate wiving by this method—arrangements such as would prepare the way for capture to pass into a form.

Of the powers of fathers in the Slav household, however, and of the constitution of the household, whether among his predecessors or his contemporaries, Nestor discloses nothing—unless there is something to be made out of a sentence which occurs in his account of Kii, the fabulous founder of the town of Kiev. The Polians, he there tells us, lived by themselves apart, each one in his own place, with their families over which they ruled. What is meant by families here, and what is meant by ruling is not defined; and thus at first sight the passage looks vague enough for every one to read his own meaning into it. From the story it appears, however, that though "ignorant people" declared that Kii was a ferryman about whose ferry a little town grew up, Nestor believed him to have been Cnaz or prince in his "family," and his "family" to have been large enough to furnish forth, or perhaps to form the nucleus of, an expedition to Constantinople. This shows clearly that by "family," Nestor, in the passage we are noticing, meant a considerable body of kinsfolk with one of the kinsfolk at their head. He narrates, too, that Kii and his two brothers at first settled each on a separate mountain—not holding together as a family or in the same House-Community; and that they afterwards united, and, coming into the plains, founded a little town which was named after the eldest brother—which is only intelligible on the view of their being chiefs and rulers over their kindred. Whether the Polian family which lived apart was one man's family, however, or a House-Community, or a clan, the passage leaves us altogether uninformed as to the powers of its head.

Of the House-Community we find nowhere any notice in Nestor. His mention of polygamy as an occasional practice among certain Slavs, makes against the existence of this family arrangement among them. Apart from this, all his descriptions perhaps are consistent with it. More than that can scarcely be said. But it may be worth adding that it is no unlikely thing that convenience at an early period established among the Slavs an arrangement which, with varying incidents, has been common among semi-barbarous and barbarous peoples.

As to the relationships acknowledged among the ancient Slavs, Nestor does not appear to have known of their varying from the system with which, as a Christian of the Greek communion, he was familiar. And it is but very rarely that his facts throw light upon them. We find one, however, which seems to exclude Agnation. The right-hand man of Vladimir the Great, in his youth while he was fighting for a kingdom, and in the administration of his kingdom afterwards, was his mother's brother, Dobrinia—who, plainly, was not only Vladimir's acknowledged relative but, his own brothers being competitors and rivals, the nearest relative he had to depend upon. Here is testimony, though of course not "legal" testimony, against the Patriarchal Theory.

Further testimony against it appears to be found in the position assigned to Olga, widow of Igor, the son of Rurik, after the death of her husband and during the minority of her son; and (since she may have been Varagian) still more in the nature of the traditions about her. Instead of living in perpetual tutelage, Olga was her son's guardian, and regent of the kingdom, with all the powers of a sovereign; and Nestor gives wonderful accounts of the revenges which—taking up the blood feud, woman though she was—she took for the murder of her husband.

We should scarcely find such a woman made a popular heroine at a time when women were considered unfit for all places of trust and condemned to life-long subjection.

As to her being regent and guardian, Sir Henry Maine, aware of the frequency of cases in which such functions have devolved upon the sovereign's mother, has accounted for them by saying that this has occurred "doubtless out of respect to the overshadowing claims of the mother."* But it is part of his theory that women had no overshadowing claims; that their lot in life was perpetual tutelage; and that the people with "overshadowing claims" were the husband's kinsmen on the father's side. The fact he had to account for, and the explanation he offers, alike show a weight allowed to natural relations which the Patriarchal Theory denies to them.

It may here be said that South Slavonian tradition freely gives the sovereign's place to a woman; and, for example, Krok, the monarch with whom Bohemian tradition begins, is succeeded by his daughter, the Queen Libusa, whom we have already encountered as the upholder of ancient custom. Sir Henry Maine, aware of the frequency of such female successions, has given an explanation of them also, and it also does not seem consistent with his theory.

The explanation, he says, "no doubt is that the circumstances of the time allowed unchecked play to respect for the claims of blood; the men being ex-

* *Ancient Law*, p. 240.

hausted, a woman was taken rather than a new strain of blood introduced." *

This explanation is simple and natural enough. But it assumes (without warrant) that women have never succeeded except when the men of the royal stock have all been exhausted; and it overlooks the fact that the succession of a woman almost necessarily introduces a new strain of blood. We are not told of the legendary Krok that he left no male relatives behind him, but that, having a daughter, she was his successor; and her succession, according to the legend, did introduce a new strain of blood, and passed a share of the government at once to a person of new blood. This by the way. What we are concerned to point out is that this explanation ignores the Patriarchal Theory. That theory does not allow "unchecked play to respect for the claims of blood." The adoption of a son would be entirely consonant with it; and it is this which would, in the case spoken of, be its natural resource—not the sovereignty of a woman. Than the sovereignty of a woman, nothing could be more antipathetic to a theory which considers a married woman as finis familiæ, and puts women throughout life under supervision.

As to the particular case of Queen Libusa, it is to be borne in mind that Libusa is undoubtedly a mythical person. She was not a hard fact that a people brought up under "Patriarchal" ideas had to make the best of. That the popular fancy invented her, and made much of her, proves that the early Slavs had no difficulty at all in

* *Early Law and Custom*, pp. 248, 249.

submitting to a female sovereign, or female chief, and even that they took kindly to one. And, indeed, if South Slavonian tradition be good for anything, the Southern Slavs were well familiarised with female headship. There could be no better proof of a people being free from the ideas which are comprised in the Patriarchal Theory. It should be said that women appear to have been in the line of succession to the throne in Russia.*

So far, the little we have gathered from Nestor—which though little is not without weight—is all against the Patriarchal Theory. We learn further from the Chronicle that the settlement of the warlike Varagians in their country soon brought the Slav populations of Russia into contact, or closer contact, with the Greeks, and that a Treaty (not noticed by the Byzantine writers, but set forth at length by Nestor) was made with the Emperor by Oleg, Rurik's successor, in the year 907. Here at last we have a chance of getting some hints as to the legal institutions of the Russians; but, so far as the present inquiry is concerned, there is scarcely anything to be made of it. Levesque inferred from the

* We find Sir Jerome Horsey, in the time of the Czar Feodor, speaking of the Czar's cousin, widow of a Duke of Holstein, as "Queen Magnus, the next heir of the emperiall crown of Muscovvia," and such she admittedly was. (Horsey was employed by Boris Godounof, who was then aspiring to the throne, to persuade her into coming with her daughter to Moscow; which having been done, both were promptly put into a nunnery.) This, notwithstanding that during nearly six hundred years, the clergy, imbued with Byzantine notions, had been steadily lowering the position of women, especially of the women of the better classes.

third article of this Treaty that, among the Russians, a wife had a share in her husband's possessions; but the text of that portion of the article on which he founds is too much in doubt for any inference to be made from it. From the tenth article he inferred that they knew the Will; but perhaps it can hardly be taken to prove more than that the Will was known to the Russians who traded to Byzantium. The former of these articles gave to the next relation of a murdered person a share of the murderer's property when he had escaped; and the latter gave the property in Greece of a Russian who died there without a Will or without having (probably wife or children) to his relations in Russia. But it does not appear who the "next relation" of the one article or the "relations" of the other were. And it cannot even be made out whether the power of willing existed when there was a family (though apparently it did not).

A second Treaty with Greece, also preserved by Nestor, was entered into by Igor, the son of Rurik, in the year 945, and in connection with this there are two things to be noted. The first is that the provision relative to murder (differing, though perhaps only in words, from that of the former Treaty) is that "the relations" of the murdered person were to have the right of killing the murderer, and of taking his property if he escaped. The other point is more important. It appears from the preamble of the Treaty that a married woman—Sphaindr, the wife of Oulieb—was one of Igor's deputies by whom the Treaty was negotiated. This Sphaindr not being either a regent or a queen, we cannot guess what explanation of her employment on such a mission can be offered by an advocate of the Patriarchal Theory.

Before the end of the tenth century, Russia, in which there had long been Christians, all of a sudden became a Christian country. Soon it swarmed with clergy, and the clergy thenceforth had an immense influence in its affairs and in the shaping of its institutions. An ordinance which assigns to them a most extensive jurisdiction, both criminal and civil, is attributed to Vladimir the Great, the first Christian Monarch, but probably it is of much later date and comprises powers of gradual growth. At any rate, the first body of law published in Russia bears no trace of Christian influence. Of course it is the more valuable on that account.

This body of law—the Laws of Jaroslav,* son of Vladimir the Great—was published in the year 1017. That it was drawn up to some extent under Scandinavian influence need not be doubted. But we can feel confident, from the nature of the matter dealt with, that the article with which it opens, which alone concerns us, embodied nothing that was not established Russian custom, and the more confident because, in its second branch, we find it conflicting with Scandinavian usage. It is, moreover, in accord with what we have gathered from Nestor. It deals with the blood feud, and names those who should have the right of taking

* Ruskaia Prawda. See *Das älteste Recht der Russen*, by J. P. G. Ewer. Dorpat, 1826.

vengeance. Christian influence must have aimed at restricting the number of such persons. A monarch, promulgating laws, would naturally seek to restrict rather than to extend it. At any rate, this article, such as it is, undoubtedly contains the best "legal testimony" relative to the Patriarchal Theory which early Russia has to offer. Its testimony is against the theory.

For it gives the right of revenging a murder to the brother, the son, the father, the brother's son, and *the sister's son* of the slain. Argument cannot be needed to show that the duty of taking vengeance has everywhere attached only to those who were in the fullest sense relations of the slain. In Russia, therefore, a man's sister's son must have been in the fullest sense his relation. This article shows the absence of Agnation.

The second branch of this article provides for the case of there being no person to take vengeance for a murder, in that case imposing a fine payable to the prince. The amount of the fine was the same whatever the rank of the murdered person, and whatever his nationality —a complete departure from Scandinavian usage.

Isiaslav, son of Jaroslav,* was also a legislator. He both modified and added to the Laws of his father, but he left the first article untouched.

It can now be understood how perplexing is the statement that important legal testimony to the Patriarchal Theory is furnished by the Slavonic peoples. Their oldest records—so far as they are good for anything—negative Agnation and, to all appearance, the

* See Ewer, cited above.

tutelage of women. Women were competent to fill the sovereign's place among the Slavs both of the North and of the South. As to the paternal power among the early Slavonians, there is simply an absence of information about it.

In the centuries following the introduction of Christianity, there were influences at work capable of effecting, and which did effect, immense changes in the social condition of Russia—the Church itself, which introduced a foreign jurisprudence, and Eastern views about the treatment of women; the Tartar domination; the incessant wars of the princes between whom the country was parcelled out; the position of autocracy finally arrived at by the Grand Prince, who thenceforth treated even the highest of his subjects as his slaves. From the boyar of the tenth century, his master's companion and counsellor, with whom he lived on terms almost of equality, to the noble of the sixteenth who, in addressing the Czar, grovelled before him, saying, "May I speak and not be whipped?" is a tremendous change, and it may be taken as a measure of what had gone on throughout the whole social system. From what is known of any Russian institution in the sixteenth century, generally speaking, no conclusion can be reached as to what that institution was in pre-Christian times. Nor will even those Russian laws which are assigned to the thirteenth century* be found helpful for the purposes of the present inquiry.

* A collection of uncertain date (the Prawda of Novgorod), but ascribed by Ewer to the period 1280–1300, contained in a volume of

Here, then, so far as *Ancient Law* is concerned, we may conclude. That work contains notices of the Russian Village Community, but Sir Henry Maine has changed his opinion as to the place of the Village Community in the history of society. Once he was disposed to identify it with the Roman Gens. He does so no longer, and so no more need here be said of it except that there appears to be little doubt that the periodical redistribution of the land among all adult males, which in Russia is its distinguishing and essential feature, was unknown in that country until after the establishment of serfage. It is not known that Village Communities (in the Russian sense) have at any time existed among the South Slavonians.

uncertain history, is given in Ewer's book already cited. These laws form a supplement to the Laws of Jaroslav and Isiaslav. It is evident that they consist of rules published at different periods, for they are not altogether consistent with one another. It is always impossible to make out satisfactorily to what extent those rules were innovations, or what it was that they replaced.

NOTE TO CHAPTER VII.

THE PATRIARCHAL THEORY AND ROYAL SUCCESSION.

If, in the default of legal testimony, or of any direct evidence, as to the social condition of the early Slavs, one were casting about for means of determining what that condition at one time was, the nomenclature of relationships, or rather the terms of address between relations, in use among the Slavonic peoples would furnish evidence of the highest importance. Simply to throw out this, and to indicate the kind of puzzle which that nomenclature presents to an upholder of the Patriarchal Theory is all that can be done here. To this day a Russian addresses his first, second, third, and fourth cousin, *whether on the father's side or on the mother's*, as brother or sister, distinguishing a first cousin from a brother or sister, however, as double-birth brother or sister, a second cousin as triple-birth brother or sister, and so on. Similar modes of speech, not confined to the case of cousins, but running through the nomenclature of relationships, occur among all the Slavonic peoples. They can only be explained as exhibiting what Mr. Morgan called the classificatory system of relationships in stages of decay. How, consistently with the Patriarchal Theory, to account for a man and his fourth cousin *on the mother's side* having called each other brothers is therefore the sort of problem which this nomenclature raises. Those who know Mr. Morgan's voluminous work (*Systems of Consanguinity and Affinity of the Human Family*) do not need to be told that similar problems are raised by the nomenclature of addresses in the case of numerous races not less important than the Slavs. It is not rash to say that a solution consistent with the Patriarchal Theory is not to be looked for.

The rule which, down to comparatively modern times, regulated the succession to the throne in Russia might also be made to throw light upon the ancient social condition of the Slavs; and it too makes a serious difficulty for the Patriarchal Theory. This rule —familiar to us as being still in force among the Turks, and which seems to have once been as well-known among Aryans as it has been among Mongols—gave the throne to the oldest male of the royal family, so that the brother of a deceased ruler succeeded in preference to his son. The Patriarchal Theory involves the succession of sons to their father—and here we find the son not succeeding his father.

Sir Henry Maine is of opinion that this rule was arrived at upon considerations of policy. The origin of it, he says, "is doubtless a simple calculation on the part of rude men in a rude society that it is better to be governed by a grown chieftain than by a child, and that the younger son is more likely to have come to maturity than any of the eldest son's descendants." * He thinks that, in general, ordinary succession and succession to the throne have had nothing to do with one another †—that the rule settled for the former, and which was doubtless deemed equitable for it, has commonly not been applied to the latter. And having made regard to the royal blood account for the

* *Ancient Law*, p. 241.

† "The King and His Successor," *Fortnightly Review*, 1882. *Early Law and Custom*, chap. v., on "Royal Succession and the Salic Law."

sovereignty of females, and respect for the claims of the mother for female regencies, he has attributed this rule to a calculation which, in some of the cases for which the rule was intended, would no doubt be a just one.

If it were admitted that rules for the succession to chiefships and sovereignties have been derived from rules for ordinary succession, the question would arise, What system of ordinary succession gives heirship not to the son but to the eldest male of the family? It would be found that this system of succession has prevailed widely. This of itself would necessarily tell against the Patriarchal Theory—and that is what we are here concerned with. But it is the system of succession which arises with Tibetan polyandry.

As for Sir Henry Maine's explanation, views of policy being apt to vary, an explanation of its kind offered for a rule which has prevailed extensively, and the policy of which is far from being absolutely good—it, though likely to work well in some cases, being likely to work badly in as many others—must always be regarded as extremely doubtful. The best that can be said for such an explanation of such a fact is that often we cannot judge whether it is right or wrong, probable or improbable. The utmost that can be allowed is that there is something to be said for it. A rule yields the results it is capable of yielding, whatever be the way in which it originated. And nearly every rule yields some good results. There would be an end of inquiry if we were generally to assume that rules were devised to accomplish that good which they effect. And we know enough to be sure that this mode of explaining is in most cases a bad one. An explanation of this kind, therefore, can in general have no hold upon belief, and cannot make any real difficulty for a competing explanation of a different kind.*

* Of course, it is quite consistent with what is here said that it should be beyond doubt that policy has had a secondary part in affect-

Now, for the explanation of rules of chiefly and royal succession which connects them with rules of ordinary succession—and which connects the Russian rule with the system of ordinary succession which gives heirship to the oldest male of the family—a very strong case can be made out. The nature of it may be indicated here.

To begin, what men have done in a certain case must outweigh any amount of speculation as to what men in that case would do. Sir Henry Maine has himself remarked—with surprise and as if it were a fault—that, in mediæval Europe, in connection with the feudal system, people tended to confound the law of succession to the throne with the law of ordinary succession. And, in fact, in connection with the feudal system, it seems never to have been doubted that if a new case arose in connection with royal succession, the law for ordinary succession must govern it; or, if an entirely new rule was needed, that this rule would apply equally to private succession and to succession to the crown. Sir Henry Maine has shown how this was illustrated in the controversy between Bruce and Baliol as to the succession to the Scottish Crown. But it was perhaps even more strikingly illustrated in another controversy which he has discussed—the controversy as to the succession to the French Crown between the collateral male heir and our Edward III. In this controversy it was never disputed between the controversialists that the law which regulated the succession to land should govern the succession to the sovereignty also. What was disputed was, whether the provision of the Salic Code as to Salic

ing chiefly succession. Whatever the rule for succession, policy might exclude the lunatic, or the imbecile, or a person infirm in body. Similarly, among the warlike Parthians, an injury to eyesight, which disqualified a man for being a warrior, is said to have disqualified him for being king—though, among the Parthians, a child might be king because he might grow up to be a warrior, and an old man of eighty remained king.

land applied either to the soil or to the Crown of France (see Shakespeare's "Henry V.," Act i., Scene 2), it being admitted that, if it applied to the one, it also applied to the other. That the provision of the Code in terms applied to land only was beyond doubt perfectly well known. What such cases prove is, that mediæval Europe could not help believing that the law of royal succession and the law of private succession should be identical, and that whatever was settled for the one was settled for the other also. Mediæval Europe bulks largely enough for us to be safe in inferring something as to human tendency from it; and its tendency was not to devise special rules upon politic considerations for the sovereignty when new rules for it were needful, but to apply to it principles which were thought good for succession in general, and, apart from cases altogether unprovided for, to apply to it the ordinary succession law. No doubt the law for the succession of chiefs and princes was well abreast of the general succession law among the Northern peoples before what we call feudalism began. Since it is not likely that the early world was less under the influence of established systems than mediæval Europe, we get at once a strong probability that when in early times a rule of succession for the chiefship came to be needed, the rule that would be applied would be the rule which was already established for the government of the family. And if the family was governed by the eldest male, the eldest male of the chiefly family would, as a matter of course, succeed to the chiefship.

Chiefly successions would be rare, however, compared with private successions—whether to the property of the family or to the government of the family. They would also be greatly more important. There would, besides, seldom be more than one or two persons interested in making any change in them, whereas a time would come when nearly all men might think themselves interested in procuring that men should be succeeded by their sons rather than by their brothers or by the sons of their sisters. And the interest—at any rate the immediate interest—of the many as regards chiefly succession would be that the settled rule should remain. On all these accounts, there would not be in the succession to chiefships the room there would be in ordinary succession for the employment of those devices by which systems may gradually be altered. What might scarcely be noticed in the latter case would be flagrant lawlessness in the former. The rules applicable to the former might, therefore, remain unchanged when the rules applicable to the latter had been radically altered. The two might differ only for a little. But until the change in private law was general and complete, a change in the rule for the chiefship could scarcely be thought of; and, even then, the persons interested in making a change would be few, and the interests opposed to it would often be powerful. The two, therefore, might differ long; and, if they did differ long, they would almost necessarily come to be regarded as independent of each other. And the longer the divorce between them lasted, the more difficult would it be to bring them together again.

That the rule for chiefly or royal succession should often differ from the rule for ordinary succession is, therefore, what is to be expected if the former was got by the application to the chiefship, when a rule for it became needful, of the rule for ordinary succession which at the time prevailed. And when we anywhere find the two differing, and that the rule for chiefly or royal succession can be identified with an old rule of ordinary succession, it is a reasonable supposition that this rule was previously in that place the rule for succession in general. In some cases it can be shown that this is the fact. And a few examples of such verification are sufficient to convert the supposition into a legitimate inference. In other cases, though verification of the supposition may not be possible, there is no alternative to it but an explanation from policy which is manifestly ridiculous. Where we find the chief succeeded by his sister's son, sons succeeding to their father in all other cases, it is indefinitely easier to believe that this happens because the sister's son was formerly the heir in ordinary succession—that form of succession law having been very common—than to think that the chiefship was exceptionally provided for to save it from a risk which all men were willing to run for themselves—with a view, that is, to ensuring that the chief should never be of other than the chiefly blood.

Sir Henry Maine has justly remarked of the rule of succession to the Sultanate—that widespread rule for which his explanation from policy has been offered—that it seems not to be derivable from the Mohammedan law of inheritance. If, however, we suppose that it has survived among the Turks from pre-Mohammedan times, the supposition can easily be verified. And, if we go on to suppose that it was once the rule for ordinary succession among them, it can at least be shown that it was once a rule of ordinary succession among the races to which they belong.

But again, if the law of royal succession lags behind the law of ordinary succession, it tends to follow it, and to become identical with it as far as the case to which it applies permits. The rule which preferred the oldest male, once so common, is now found among no people of any importance except the Turks; and the natural wish of fathers that their sons should succeed them in the throne as in private successions has recently endangered it as a rule even among the Turks. A mode of nullifying it—fratricide—has long been in use among them, as it is among many barbarous peoples who have the same rule. In small tribal communities a form of election has sometimes appeared with it in its decay; and this has doubtless been a device for averting the disputes about succession which arose when its justice and policy had fallen into controversy. This expedient gave the chief's son a chance of succeeding him, and it enabled the tribe to settle in each case whether the old rule should prevail or not. Of course every departure from it which was thus sanctioned helped to break it down as a rule, and so helped to prepare the way for the succession of son to father. Another rule of chiefly succession which has been mentioned, that which gives the chiefship to a sister's son, appears to have been nullified in some cases by means of an extraordinary but effective expedient—by the chief, that is, marrying his own sister.

That the tendency of mankind to apply to a new case the system they have already for similar matters made the earliest rules for chiefly succession and the contemporary rules for succession to the government or to the property of the family identical—a view corroborated by our finding that rules for chiefly or royal succession are always, as nearly as circumstances permit, and saving what are obviously expedients of transition, identical with known rules of ordinary succession—identical frequently with the current rule of ordinary succession, and always tending to approximate to it; identical in other cases with a more primitive rule which can occasionally be shown to have preceded the current rule in ordinary succession—seems to give a sufficient explanation of the rule which regulated succession to the crown in Russia. In judging of the sufficiency of this explanation it is difficult not to think of the alternative one. No doubt rules for chiefly succession formed in many different places, under every variety of circumstances, upon considerations of policy, might happen all to coincide, and might also happen all to coincide with a discarded rule of ordinary succession. But that this should have happened cannot but be deemed very extraordinary.

CHAPTER VIII.

UNDIVIDED FAMILIES AND HOUSE-COMMUNITIES.

It now becomes necessary, by way of supplement to the two preceding chapters, to consider how the existence of the Joint Undivided Family among the Hindoos, and of the House-Community among the Slavonians, bears upon Patria Potestas.

And, to begin, let it be said that we know nothing at all of the latter, and very little of the former, except from comparatively recent descriptions. From the early Hindoo writings it might be a justifiable inference that undivided families were very much rarer and smaller in dimensions among early Hindoos than they are now. That, perhaps, is not a matter to dogmatise about; but it is certain that those writings tell us extremely little about the Undivided Family. And their teaching was anything but favourable to it, for they commonly inculcate upon fathers that they should divide the family estate in their lifetime. Of Village Communities, it may be said, those writings tell us literally nothing. There is nothing in them from which it can be gathered even that such communities were known.

Our knowledge of the Slavonian House-Community, again, whatever may be the antiquity of that institution, is entirely derived from modern times. Its existence in pre-Christian times, though anything but improbable, is matter of conjecture; and what it was then nobody knows. It is from accounts of it written after the Slavonians had been for many hundred years Christians, and more or less in contact with other races, that such inferences as can be made from it have to be made. The Village Community is not known to have ever existed among the South Slavonians.

It does not seem reasonable to think that conclusions as to the primitive social condition of man are to be obtained from what we know of either of those institutions.

It should be said that Undivided Families, or House-Communities, or, to speak more generally, households composed of the persons nearest to each other in acknowledged kinship, are not found with one family system only, with one system of kinship or inheritance only. They are constantly found among peoples whose material condition is not much advanced, whatever their family system. But it would seem that if they are more proper to one family system than to another it is to the family system founded upon kinship through females only. At any rate, as Mr. J. D. Mayne has pointed out,* "the most perfect form of the joint family now existing" is found among the Nairs of Malabar, with kinship acknowledged through females only, and

* *Hindu Law and Usage*, pp. 179 and 192.

paternity unrecognised. A people which in the nineteenth century has House-Communities composed chiefly or entirely of persons related to each other through male descents may possibly have had, at an earlier time, House-Communities composed of persons related to each other through female descents. At any rate, all that can be inferred from their now living in House-Communities (that they acknowledge kinship through males being otherwise known) is that their material condition is such that this way of living still suits them. And we must go to other sources when we seek the means of concluding whether they have always acknowledged relationship through male descents or not. As to this, the present constitution of their House-Communities cannot prove anything.

So much having been premised, we go on to state Sir Henry Maine's view of the relation of the Joint Undivided Family and of the House-Community to Patria Potestas. It can be stated very briefly.

He tells us that the Joint Undivided Family,* "wherever its beginning is seen" in Aryan Communities, springs universally out of the Patriarchal Family, with despotic powers centred in the Paterfamilias; and that, as in such communities we find it springing from a Patriarchal cell, so, when it dissolves, we see it dissolving into a number of such cells. This he says equally of the House-Community of the South Slavonians and of the Joint Undivided Family of the Hindoos; for he holds the two to be identical, and both to be

* *Early History of Institutions*, pp. 116–118.

identical with the Gens (or alternatively with the body of Agnates) of the Romans. The Hindoo and Slavonian cases are the only cases of this kind he has considered.

Now, as to the former, it has already been shown that evidence of Patria Potestas having at any time prevailed among the Hindoos is altogether wanting; and especially that the "legal testimony," to which alone *Ancient Law* made appeal, proves not the prevalence of it, but the absence of it. Among the Hindoos, as a fact, we do not see the Joint Undivided Family either springing from a Patriarchal cell or dissolving into a number of such cells. We only know it as springing from, and also as succeeded by, a family system in which the paternal power was severely restrained, and in which wives and children had rights inconsistent with the Patriarchal Theory.

But is there not in the constitution of the Joint Undivided Family among the Hindoos something that makes so strongly for Patria Potestas as to impress upon us irresistibly that there is a close connection between the two? No. Sir Henry Maine himself tells us that the head of such a family is not a Paterfamilias; that he is not the "owner of the family property, but the manager of its affairs and the administrator of its possessions;" and as to the Power of the Father, the utmost he can say is that each father or grandfather "has more power than anybody else" over his wife and his descendants. He allows, in fact, that there is nothing at all approaching to, or suggestive of, Patria Potestas to be found in the Hindoo Joint Undivided

Family. The head of the household is only an administrator; the father only has (and it would be strange indeed if less could be said of him) "more power than anybody else" over his family.

One could understand it being inferred that, wherever the Joint Undivided Family has been of common occurrence, Patria Potestas must have been unknown. As has been said already, the most perfect specimens of the Joint Family are found where fatherhood is not recognised.

Passing now to the House-Community of the Slavonians, it has to be recalled that, while a good deal of evidence casting doubt on the Patriarchal Theory has come from Slavonic sources, in the scantiness of our knowledge of the early Slavs, we have no information as to what the paternal power was among them. Sir Henry Maine, being without means of showing what was the constitution of the family out of which the House-Community originally sprung among the Slavs, has been content with inquiring into the constitution of the family which issues from it now—that is, of the ordinary family, or, as he calls it, the "natural family," now found, and quite commonly found, among the South Slavonians. On his theory, the House-Community must have sprung out of the Patriarchal Family, and that in turn should issue out of the House-Community. Accordingly he has given reasons on which he thinks it may be concluded that the natural family now found among the Southern Slavs is a Patriarchal family "despotically governed by the oldest ascendant." He has arrived at this conclusion, however, by inference. And he has put aside testimony which makes that way of getting a conclusion unnecessary.

He gives two reasons for his inference. The first is that there is in South Slavonian countries a great respect for old age. The other is that in these countries a distinction is maintained between agnatic and cognatic relationship.

As to the former, respect for old age is common in nearly every country, whatever the degree of parental authority or the system of kinship; and it has nowhere been stronger than in Red Indian villages in which, relationship being counted only through women, a son was not even a relative of his father. Add to this, that, among the South Slavonians, it extends equally to old women and old men, and there need be no hesitation about declaring that the respect for age shown by the South Slavonians is absolutely irrelevant to an inquiry as to the powers of fathers among them.

As to the second, the fact founded upon is that, among the South Slavonians, relations through male descents are called relations of the great blood, and relations through female descents relations of the little blood. This shows, no doubt, some preference for relations through male descents. But it cannot show whether formerly that preference was greater than it is now or less; and still less can it show that at one time relations through female descents were not acknowledged as relations at all. A preference for relations through male descents, that is, is not Agnation, and cannot prove that relationship was formerly agnatic. And among the South Slavonians, in point of fact, relationship through women is most fully acknowledged and carries rights of succession—whereas Agnation denies all relationship and, above all, all rights of succession to persons claiming through women. Sir Henry Maine, however, has concluded that he has here found a recognition of agnatic relationship. Applying his formula, that this is good evidence of Patriarchal Power subsisting or having formerly subsisted, he then infers that the natural families of the South Slavonians must be Patriarchal families "despotically governed by the oldest ascendant." Let us now inquire what they actually are.*

Though families of an ancient type, they are extremely unlike the Patriarchal Family of Sir Henry Maine's theory. The man, no doubt, is the head of the house, and the wife is expected to "honour and obey" him. But she is not without honour and influence in her turn. She is absolute mistress in domestic matters. She always has separate property, more or less, and transmits it to her children. As regards the children, the rights and duties of the spouses are the

* Such preference for relations through male descents as is found among the South Slavonians, and the phrases which express it, appear to be fully accounted for by the fact that such relations and their children habitually live together with common interests in the same House-Community; while their relations through women are in other House-Communities in which they have common interests with strangers. If this be a true account of the preference, it has nothing to do with Patria Potestas, and is among the results in working of a family arrangement under which—as will be seen by-and-by—parental power is curiously restricted.

same; and if the father dies, the mother takes his place till the eldest son is old enough to manage. The mother naturally has most to do with the girls and little ones, the father, though over all, taking for his special charge the sons as they grow old enough to be useful. So far, these households exhibit nothing unfamiliar; nothing very different from what, with the inevitable divergences between theory and practice, varying from country to country, from house to house, is to be found in other parts of Europe. There are, nevertheless, between these and Western households remarkable differences—some of which it is important for us to note. In some parts of South Slavonia, children are allowed to keep certain of their acquisitions for themselves, but commonly among the South Slavonians—especially among the poorer folk—so long as they remain at home, they hand over all their earnings to their father, who, on his part, provides for all the wants of the family. On the other hand, a son on marrying is free to leave his father—who thenceforth has no control over him or his affairs; and, when he does leave, he carries his proportional share of the family possessions along with him. And, though it is at marriage that this is usually done, the custom of the country allows every grown-up son, whether married or not, at any time to go forth from the paternal household, taking with him his share of the family possessions. Daughters, it should be said, are provided for by dowry, and only become their father's heirs (which they are in that case though married) when there are no brothers. Sometimes a whole household agrees to divide and break up; and then the son with whom the old people elect to live—who is usually the youngest—gets on account of that a somewhat larger share than that to which he would otherwise be entitled. Enough has now been said to distinguish the Slavonian from the Western family, and enough to show that it differs essentially from the Patriarchal Family.*

Both wives and children have rights which the Patriarchal Theory denies them. The wife has property; and, in succession to her husband, she may become the head of the house. Failing brothers, a married daughter may be her father's heir. The son, instead of being a slave, has rights which, in families of a more modern type, sons do not enjoy. The father, no doubt, has that power over the young and helpless which, subject to the limitations of law and opinion, parents can everywhere exercise. But that is not Patria Potestas. He keeps the common purse and manages while his children are young—there is nothing very strange in that. What is strange is that his sons, as they grow up, can leave him one by one, each stripping him when he does so of a portion of what in other countries would be accounted his possessions. That his son, once grown-up, should be free to assert independence of him would be contradiction sufficient of the Patriarchal Theory. That he should be then entitled both to independence and

* See Fedor Demelic, *Le Droit Coutumier des Slaves Méridionaux. D'après les Recherches de M. V. Bogichich.* Paris, 1877.

his share of the family possessions discloses a view of family right which is separated by a gulf from that view of it on which the "oldest ascendant" is lifelong lord of his family and its property. But it is apparent that in every important relation the Slavonian natural family discountenances the Patriarchal Theory. It shows, what a multitude of examples show, that a family may be of ancient type, and yet be extremely unlike that Roman family which Sir Henry Maine has called the Patriarchal.

When we come to inquire whether there is anything in the constitution of the House-Community that makes for Patria Potestas, the answer must be, as it was in the case of the Joint Undivided Family of the Hindoos, that there is nothing; and that it is difficult to believe that a high view of paternal power could be entertained in a country in which such communities have been very popular for ages. Slavonic writers have insisted upon the dissimilarity between the House-Communities and the Patriarchal Family, and upon the inapplicability to the former of such words as patriarchal. M. Fedor Demelic, from whose little tractate Sir Henry Maine has chiefly derived his knowledge of those institutions, has fully recognised the justice of this protest; and so, indeed, has Sir Henry Maine. Briefly, the House-Communities are, among the South Slavonians, industrial partnerships—formed, no doubt, among relatives—in the management of which, in general, all the grown men and all the married women have a voice; in which a father and every grown-up son of his have precisely the same position and rights; having at their head an elected manager who, though much controlled, has considerable powers for management and discipline, to which all, young and old, are subject, and which, therefore, cross and cut down a father's control over even his younger children. They are, therefore, societies in which the Patriarchal Theory is set at open defiance. It would seem to be highly improbable that, immediately out of a family in which the "oldest ascendant" was supreme, there ever issued a society in which he has only equal rights with his grown-up son or grandson. This would be despotism giving birth to democracy. That the "oldest ascendant" should—unless he happen to be the elected head—have only equal rights with his descendants can in no possible way be made to support the theory that he was at a former time supreme over them all.

It will be proper, however, to show somewhat fully what the South Slavonian House-Communities are, and it will be well, by way of preface, to give some account of the peasant household as it is (or till lately was) in Russia.

Before the emancipation of the serfs, the peasant household in Russia, Mr. Wallace tells us,* often contained representatives of three generations; all, young and old, living together under the direction and authority of the head of the house, called usually Khozain or administrator, or in some districts Bolshak, "the Big

* *Russia.* By D. MacKenzie Wallace. Cassell, Petter and Galpin: London, 1877. Vol. I., pp. 134–7.

One." The Khozain's position was, generally speaking, occupied by the grandfather or, if he was dead, by the eldest brother; but this rule was not strictly observed. If the grandfather was infirm, or the eldest brother incapable, the place of authority was taken by some other member—it might be by a woman—who was a good manager and possessed the greatest moral influence. The relations between the head of the house and the other members depended upon custom and personal character, and they consequently varied greatly in different families. "The house with its appurtenances, the cattle, the agricultural implements, the grain and other products—in a word, the house and nearly everything it contained—was the joint property of the family. Hence nothing was bought or sold by any member—not even by the Big One himself, unless he possessed an unusual amount of authority—without the express or tacit consent of the other grown-up males, and all the money that was earned was put into the common purse." "The peasant household of the old type," Mr. Wallace proceeds, "is thus a primitive labour association, of which the members have all things in common, and it is not a little remarkable that the peasant conceives it as such rather than as a family. This is shown by the customary terminology and by the law of inheritance. The head of the household is not called by any name corresponding to Paterfamilias, but is termed Khozain or administrator—a word that is applied equally to a farmer, a shop-keeper, or the head of an industrial undertaking, and does not at all convey the idea of blood relationship.

The law of inheritance is likewise based on this conception. When a household is broken up, the degree of blood relationship is not taken into consideration in the distribution of the property. All the adult male members share equally. Illegitimate and adopted sons, if they have contributed their share of labour, have the same rights as the sons born in lawful wedlock. The married daughter, on the contrary—being regarded as belonging to her husband's family—and the son who has previously separated himself from the household, are excluded from the succession. Strictly speaking, there is no succession or inheritance whatever, except as regards the wearing apparel and any little personal effects of a similar kind. . . . The members do not inherit, but merely appropriate individually what they had hitherto possessed collectively. Thus, there is properly no inheritance or succession, but simply liquidation and distribution of the property among the members."

This statement of an excellent authority leaves little need for comment. It is plain that the oldest, if fit, was always the head of the family, but that capacity was too important to be long dispensed with, even in the case of the common progenitor. And a woman might, as the most capable, come to have the management. As to the powers of the manager, they were what his personal character and the circumstances of the family happened to procure him. But he no more had the powers of a Paterfamilias than he had the name. As regards control over property, indeed, he was so very far from having the position of a Paterfamilias that he could do nothing involving gain or loss —could neither sell nor buy—without the consent of the other members; while he got, if from any cause the household broke up, simply an equal share with the others. When the family consisted, as it often did, of one man and his descendants, it was identical in composition with the Patriarchal Family. Observe then the disparity between the powers ascribed to the head in the one case and the powers actually possessed by the head in the other. The Russian "Patriarch," instead of being master in all things, governing all despotically, stood in so different a relation to his household that he was called administrator or manager; and while, in other respects, he had such power as he could assert for himself, he needed for his actings, even the most trifling, affecting property, the consent of his grown-up descendants—they being, as his co-partners, equally interested with himself. Here, as in the South Slavonian natural families, we find prevailing a view of family right utterly different from that which underlies the Patriarchal Theory. And the rule of inheritance, or rather of partitioning, of these Russian households, as explained by Mr. Wallace, clearly gives the suggestion that married women and forisfamiliated sons were cut off from succession, not on any profound theory of the family, not for example through any fear of "society being confounded," but on the simple ground that those who had been conjoined in producing property, and were the joint owners of it, were the persons who should keep or divide it.*

The South Slavonian House-Community also is regarded as an industrial association. And that is precisely what it is. Sometimes it springs out of the "natural family," through the members of such a family, notwithstanding that they are free to separate, electing to remain in association after the death of their common progenitor; and, in that case, in its earliest stages it must closely resemble the Russian peasant household which has just been described. But it appears very commonly to spring out of the ruins of an older House-Community—persons who had been associated in a community which has been dissolved recombining themselves, at their own choice, in a number of new ones. The House-Community, which is never large—including on an average, according to

* The House-Community in Russia was kept alive, and the authority of the chief as much as possible upheld, by the land-owners, for reasons of their own, during the existence of serfage. Since the emancipation of the serfs it has rapidly been disappearing. It has already been stated that evidence of its antiquity seems not to be forthcoming. Even the travellers of the sixteenth century have not noticed it. Fletcher and Horsey appear not to have thought of the Russian family as in any way different from the English; and the former tells us that "the commons pass over their lands by discent of inheritance to whichever son they will, which they commonly do after our gavel-kind, and dispose of their goods by gift or testament without any controlment."—Fletcher's "Russe Commonwealth" in *Russia at the Close of the Sixteenth Century*. London, 1856. Printed for the Hakluyt Society, p. 27.

But the more ancient the House-Community, the deeper does the Patriarchal Family seem to be relegated into times of darkness.

M. Bogichich, from twenty to twenty-five persons—which never is large enough to form even a small village, though all the inhabitants of a village are occasionally of the same name—comes after a time to be of inconvenient size, and then it splits up. Thereupon, while some of the members may go to live apart, each with his own family, the majority reunite at their pleasure in new communities. These are obviously voluntary associations, founded upon express contract, though subject to conditions which custom has established. And though they consist of persons who have previously been associated together, and are therefore all or nearly all of the same stock—while both convenience and family feeling tend to keep together those who are most nearly related to each other—they do not necessarily consist of near relations.

Whether the House-Community originates in the one way or in the other, the object of association is the same. The Community strives to provide for the families contained within it, not everything which their standard of comfort requires, but necessaries—lodging, food, and clothing—according to a customary not over-lavish scale. While, therefore, there is combination in labour, and a common purse among the associates, and while much care is taken to secure that everyone does his duty by the Community, there is need for separate industry. And there is also room for private accumulations. Each grown-up man works as he can for himself, to supplement what the Community gives him. His wife and, when they are young, his children work for him. And what is saved of the earnings of such industry is his own. What comes to him by inheritance also is his own. Moreover, the married woman has her dowry reserved as her separate property; and, as a rule, even the younger folks, though most of their earnings go to their father, are allowed to make their little hoards.

The Community owns the land tilled in common and the buildings upon it, the stock and the instruments of labour, the products of the joint industry, and the savings made out of them. A council, composed of all the grown-up men and the married women (the latter being in most places free both to speak and to vote) considers and decides upon every matter of any importance affecting the common interest—unanimity of decision being required in some districts; in which case, of course, a single person, if obstinate, can defeat the wish of all his associates. And at the head of the Community is an official called the Domacin, who distributes to the men their daily tasks and maintains discipline; who buys and sells—not, however, except in trifling matters, without the consent of the council—and who is, generally speaking, the representative of the Community in its dealings with those outside it. There is a Domacica too (usually the Domacin's wife), who has similar powers and duties as regards the women. The Domacin is elected, though, where one person is from age and capacity marked out in the opinion of all for the place, the form of election may be dispensed with; and he may be deposed, but the consent of all being necessary for this, he usually holds office till age makes him unfit for it. He is generally one of the seniors, and —at any rate where all are near relations—the oldest, if capable, seems to be preferred; but capacity is indispensable, and therefore the choice of the household frequently falls upon one of the younger men, and sometimes even upon a married woman. His authority varies with the trust reposed in him; but he always has the considerable powers which are deemed necessary to secure the just performance of work and to maintain order; and these, while they extend over all, are specially exercised over all the males, young and old alike, who are fit for labour—children of tender years only being left to the control of their parents. The Domacin can admonish any member; he can chastise none, unless it be the very young people. It is in dealings with property that he is most under control, but he is accountable to the council in all things; and, indeed, with such powers as are thought indispensable for his office, he is only *primus inter pares*, holding his place on condition of good behaviour. In a partition of the common goods he gets only the same share as any other, with, in some places, a present of some value added. For such partitions there is no uniform rule among the South Slavonians—the distribution being made in some districts *per stirpes* and in others *per capita.* But excepting that in Herzegovina, and probably in other places, young women who have no brothers share like men, and that everywhere some provision is made for the helpless members of the household who are not entitled to share, it is the adult males only who divide the common property between them.

This account of the South Slavonian House-Community, very imperfect as no doubt it is, suffices to make clear the points that are material for the present inquiry. Though composed of relatives, the persons who form it need not be each other's near relations. It is usually not a family, and it is a partnership—the partners being all the adult males. But the married women are consulted in everything, and a married woman may be the head of the association, with control over all the men belonging to it. A young man, once come to manhood, has in this association the same position and rights in all respects as his father or grandfather; and a father's control over his son is, between a very early age and manhood, in a great measure superseded by that of the Domacin. No man has ampler rights than the House-Community allows to grown-up sons. And a father's power over even young children is greatly restricted. This society undoubtedly has a lively sense of the uses of authority. But it makes unusually little of the Power of the Father. The powers of the Domacin, again, considerable as they are, are all given to him for the sake of good management. He possesses them in virtue of election; and he holds them at the pleasure of his associates—to whom his relation is simply that of an equal whom their own choice has set over them and invested with managerial powers. As to the women, a woman may be Domacin. And the married women sit with the men in the council which controls the Domacin. In this council, the husband, the wife, and their adult son—and no doubt often grandfather and grandson—meet each other and their neighbours for consultation on the common interest with equal rights. If the opinions of the old have more weight than those of the young, that is only what usually happens in all assemblies. But in some districts, as has been said, any one—even the youngest —of the associates has it in his power to thwart a project though all the others, Domacin included, are bent upon it.

The South Slavonian House-Community, therefore, is not so much a family as an association of persons who are relatives, and who need not be near relatives— an association quite commonly formed by persons who have previously been associated together along with others, and whose interests can conveniently be conjoined. And, as has been said already, it is in all cases an association which at every point sets "Patriarchal" ideas at defiance. The family which issues out of it, or is found alongside of it, too, is utterly different from the Patriarchal Family. Paternal power is so far in suspense in this House-Community that we should be prepared to say of it, even more than of the Hindoo Undivided Family, that the people among whom it has been popular could not have had, or could have cared nothing about such paternal powers as are denoted by Patria Potestas. Economical considerations govern it, and have taken from fathers not a little of the power which modern societies allow to fathers, in order that the powers necessary for control and management may be possessed by an elected head—whose powers, nevertheless, differ both in kind and in degree from those of a Paterfamilias. And so greatly have the House-Communities affected current notions that the relation of the head of an ordinary family even to his children is that of a Domacin rather than that of a father.

It would be unreasonable to make inferences from the South Slavonian House-Communities against the Patriarchal Theory. But it would be even less allowable, were it possible to do so—which clearly it is not—to make inferences from them in its favour.

It is obvious that, though Communities of a natural type arise, on the whole the House-Communities have drifted from the natural type of the Joint Family or House-Community—which we find in India, in perfection among the Nairs, with unrecognised fatherhood, and more commonly, in a less stable form, based upon kinship through male descents. It commonly happens that those relations whom it suits to do so—and they are usually persons who have already been associated together—club their resources; and the result is a House-Community, in which partnership ideas predominate, in which all adult men are equal, and in the management of which the women have their full share. The South Slavonian House-Communities have all grown out of a natural type of the Joint Family, no doubt. They are not commonly identical in composition with the Hindoo Joint Family, but they sometimes are; and it may be believed that at some former time they commonly resembled it more or less. They would prove, if proof were needed, that joint families can exist among people who acknowledge kinship through male descents. And they would prove, if proof were needed, that kinship through males has long been acknowledged among the South Slavonians. But they cannot inform us what paternal power, or what limitation of kinship, accompanied kinship through males among the early Slavonians; or whether that was the earliest kinship acknowledged among them; or even on what system of descents the earliest House-Communities were based among them. For views about such matters—unless we choose to do without them—we must go to other sources. The same might be said of the Joint Family of the Hindoos, but it is not necessary to say it. We know a good deal about the early condition of the family among the Hindoos; and it suffices to repeat that we do not find for it at any time any proof of Patria Potestas.

The South Slavonian Communities, short-lived as they usually are, cannot even give us suggestions (as Sir Henry Maine thinks they can) as to the manner in which chiefly authority arose. For such suggestions we must look to communities which have produced chiefs, not to communities which, so far as we know, have never produced them. And, indeed, the history of the growth of Power may be as hopefully studied in the directorate of a joint-stock bank or of a railway company as in a South Slavonian House-Community.

Sir Henry Maine at one time identified the Roman

Gens with the Village Community. He has since that identified it with the Joint Family and with the House-Community.* And in his latest work† he has given us the alternative of identifying these with the Agnates of Roman relationship. Between views so fluctuating a selection must here be made; and only a few words can be said of his identification of the Joint Family and House-Community with the Gens. What he says (or rather what he formerly said) is that there can be no reasonable doubt "that the House-Community is the Gens;" and that the Gens "actually survives" in the Joint Family.

The Joint Family and House-Community grow up within the limits of known relationship. They include at most only a section of the Agnates. And they do not last—for which the reason is obvious. After a time they become inconveniently large, and then they break up. Convenience leads to their being established, and favours them for a period, but it ultimately destroys them. And, when they have been dissolved, they leave no mark upon the future relations of those who have been associated in them and their descendants—though, of course, their relationship is unaffected. The people who compose them, among both Hindoos and South Slavonians, acknowledge—and their predecessors, so far as we know, have always acknowledged—relationship both through male descents and through female descents; and are themselves but a section, often a very

* *The Nineteenth Century*, December, 1877, p. 799.

† *Early Law and Custom*, p. 239.

small one, of those who are related to each other through male descents—being, in fact, relations who have interests which are common, or interests which can conveniently be conjoined. Both Hindoos and South Slavonians, too, acknowledge, as the Romans did, a wider relationship than can be traced, a relationship which has left marks upon usage. The Hindoos have the Gotra, which includes all persons who bear the same family name, and marriage was forbidden between persons of the same Gotra. Similarly, the South Slavonians have the Clan (as some of them have it still, it may be presumed they all had it formerly), consisting of persons of the same name, and marriage is forbidden within the Clan.

Now, the Roman Gens consisted of persons who bore the same family name, and who had only a tradition of relationship. And there is some reason to think that marriage between its members was anciently forbidden. It appears to correspond to the Gotra of the Hindoos, and to the Clan of the South Slavonians, and not to the Joint Family or the House-Community.

It is clear, too, that Joint Families or House-Communities cannot even throw light upon the early history of Gotra, Clan, or Gens unless the Patriarchal Theory is assumed.

CHAPTER IX.

PATRIA POTESTAS IN IRELAND.

In his *Early History of Institutions* (p. 216), in a passage already quoted, in discussing what he calls a perplexing problem relative to the Family among the early Irish, Sir Henry Maine says he thinks he can assign "some at least plausible reasons" for believing that the organisation therein involved was "a monument of that Power of the Father which is the first and greatest landmark in the course of legal history." He has found corroboration of this view in the fact that the father among the Irish had (along with various other persons) "judgment and proof and witness" over his son—the power, if it was a power, thus enigmatically referred to appearing to him to be Patria Potestas. Of his solution of the problem he did not speak over-confidently in the first instance. But he has since been quite confident in asserting that he has found Patria Potestas among the Irish.

The problem was raised in the first instance by a passage in the Book of Aicill,* under the heading,

* *Ancient Laws of Ireland*, Vol. III.

"What is the Reciprocal Right among Families?" and what it is may in a general way be gathered from the following passage, taken from the preface to the third volume of the *Senchus Mor*: "The most remarkable custom described in the 'Book of Aicill' is the fourfold distribution of the family into the 'geilfine,' 'deirbfine,' 'iarfine,' and 'indfine' divisions. . . . Within the family seventeen members were organised in four divisions, of which the junior class, known as the geilfine division, consisted of five persons; the 'deirbfine,' the second in order, the 'iarfine,' the third in order, and the 'indfine,' the senior of all, consisted respectively of four persons. The whole organisation consisted, and could only consist, of seventeen members. If any person was born into the 'geilfine' division, its eldest member was promoted into the 'deirbfine;' the eldest member of the 'deirbfine' passed into the 'iarfine;' the eldest member of the 'iarfine' moved into the 'indfine;' and the eldest member of the 'indfine' passed out of the organisation altogether. It would appear that this transition from a lower to a higher grade took place upon the introduction of a new member into the 'geilfine' division, and therefore depended upon the introduction of new members, and not upon the death of the seniors."

Only men, or male persons, however, could be included in the divisions, and the organisation might consist of fewer, though it could not include more, than seventeen persons. Divisions, too, are spoken of as if they regularly existed in a group of four related to one another. The passage from which we learn that, on a new birth into the geilfine division, its numbers being complete, the oldest member of that division had to move into a higher division, assumes that the four divisions would usually be in existence when such a transference took place, and that the removal of the oldest member from each of the three other divisions would follow upon it. And when two of the four divisions became extinct through the death of their members, there was a provision made for new divisions being formed in place of them—as if to restore the organisation to its normal state. It is laid down that, in this case, the surviving divisions were not to inherit the property of the divisions which were defunct, unless they were able to form and did form, out of the family, two divisions in place of those which had failed. The case of the geilfine and indfine divisions having become extinct is one actually put, and it is assumed that there would be at hand members of the family as fit to be placed in a new geilfine and a new indfine division as the persons who had died. It should be added that it is mainly through the laws of succession which connected each of the divisions with all the others that we know—so far as we do know—about the organisation. When all the members of any of the divisions were dead, its property was divided between the other divisions, and the deirbfine division, the third in seniority, was the most favoured as regards successions; the iarfine coming next, and the geilfine, the youngest division, in the third place, while the oldest division, the indfine, fared worst of all.

The composition of this family organisation and its purposes are discussed at length in *Studies in Ancient History* (pp. 453 *et seq.*). But at least five different theories of its composition have been propounded. And it is anything but strange that there should be a conflict of views on that subject. For the notices which we have of this organisation are scanty and fragmentary, while their meaning is often a puzzle in the English version, and must have been still more of a puzzle in the original. Enough can clearly be made out about it, however, to enable us to judge whether Sir Henry Maine is right in regarding it as "a monument of the Power of the Father." It is only so far as they are necessary for that, that the facts relating to it concern us here. And now nothing need be added but a passage from the Book of Aicill, which undoubtedly throws light on the composition of the organisation, and which has suggested to Sir Henry Maine a portion of his theory. The passage runs: "If the father be alive, and has two sons, and each of these sons has a family of the full number, *i.e.*, four, it is the opinion of the lawyers that the father would claim a man's share in every family of them, and that in this case they form [literally, there are] two geilfine divisions; and if the property has come from another place, from a family outside, and there be in the family a son or brother of the man from whom it has come, he shall not get it more than every other man of the family." This passage shows that lawyers contemplated as possible that *two* of the organisations might be formed among one man's descendants, and it gives their view of what the man's rights of property would be in that seemingly unusual case.*

Sir Henry Maine's theory is that any man of a sept might become a root from which might spring as many of the groups of seventeen persons as he had sons. "As soon as any one of the sons had four children, a full geilfine sub-group of five persons was formed; but any fresh birth of a male child to this son, or to any of his male descendants, had the effect of sending up the eldest member of the geilfine sub-group, *provided always he were not the person from whom it had sprung*, into the deirbfine. A succession of such births completed in time the deirbfine division, and went on to form the iarfine and the indfine." On this view, the fifth person in the geilfine sub-group was the man who, with his descendants, made up the organisation. He always remained in that sub-group, and was the "geilfine chief" or Paterfamilias. The geilfine sub-group was his hand-family, or Family under Power; while the other groups were made up of the emancipated descendants, "diminishing in dignity in proportion to their distance from the group (the geilfine) which, according to archaic notions, constitutes the true or representative family." Sir Henry Maine has not attempted to give any account of the purposes for

* The case put may have been a hypothetical one. "The lawyers" are much given to pronouncing upon hypothetical cases; and the passage reads as if the law laid down was laid down speculatively rather than known from actual practice.

which the organisation existed, nor to show its relation to Patria Potestas.

The objections to his theory are overwhelming, and there are several, each of which must be fatal to it.

1. Sir Henry is obliged to assume that the eldest member of the geilfine division (whom he takes to have been the father) never left that division. But this is contrary to what is expressly stated—which is that, the geilfine division being full, on a new member being born into it, the *eldest* member was promoted into the deirbfine.

2. On Sir Henry's theory, the extinction of the geilfine division would mean that the Paterfamilias and all his family under power were dead—that the family was extinct. But the extinction of the geilfine division did not put an end to this Irish organisation. The three other divisions divided its property, and the organisation went on. Moreover, as has been said, an extinct geilfine division could be replaced. And when another division besides it was extinct, it and that other *had* to be replaced as a condition of the organisation not collapsing. But a Paterfamilias and his family under power would have been irreplaceable. A Paterfamilias and his family under power, therefore, cannot have constituted the geilfine division, and the organisation cannot have consisted of a Paterfamilias and his descendants.

3. But further, there is no authority for conceiving that all the members of the other divisions had originally been in the geilfine division. There is only

one case mentioned in which men passed out of one division into another. That was when, with the organisation complete, a new, that is a sixth person, was born into the geilfine division. Then the senior member of each of the three first divisions was, as the editors of the *Senchus Mor* say, promoted, and the senior member of the oldest division had to go out.

Again, while the geilfine was undoubtedly the youngest division, there seems to be no authority to show (what is essential to Sir Henry Maine's explanation) that it was the first in dignity, and that the others diminished in dignity with their remoteness from it.* If the right of inheriting divisional property be a gauge of dignity, it was inferior to two of the other divisions, the deirbfine and the iarfine divisions both faring better than it. This, at any rate, was the unlucrative result of its being the youngest division. When any division became extinct by the death of its members, its property went to the other divisions in definite shares; and the rule of sharing, which preferred the junior to the senior, and the nearer junior or senior to the more remote, operated most favourably for the two middle divisions.

So far as to the incompatibility between Sir Henry Maine's theory of the organisation and the facts disclosed respecting it. It remains briefly to point out

* See *Ancient Laws of Ireland*, Vol. IV., p. 207: "After this the indfine land goes before the deirbfine land, for, in tribe lands, the indfine land is the land in which it (the water) is detained *in a pond;* for the land in which it is detained is nobler with the Feini than the land out of which it is drawn."

some discrepancies between his theory and Patria Potestas.

1. Taking the case put in the Book of Aicill upon which he has founded, it will be seen that for the father whom he regards as a root from which might spring as many of the organisations as he had sons—for the father who, having two sons with four children each, might, according to the lawyers, set on foot two geilfine divisions—he can find no place within the organisations sprung from him. His explanation, that is, shuts the true Paterfamilias out of the family. Of the violence herein involved to the principle of Patria Potestas he has himself remarked that it has no analogy in Roman Law.

In the organisations as they really were, however, the "oldest ascendant" would, in the case supposed, have been the most important person—at any rate as regards property; for, in the opinion of the lawyers, he would have been entitled to share with every family derived from him, as if he were included in it. And the words suggest that there would have been more families than two.

2. Emancipation from Patria Potestas disinherited the emancipated. But our knowledge of the divisions of the Irish family is mainly derived from the laws of succession to property which connected each of them with all the others.

3. Emancipation could only account for there being two classes of descendants, the emancipated and the unemancipated. Here we have, according to Sir Henry

Maine, four classes of descendants. Of this also it may be said that it has no analogy in Roman Law.

4. What power was it that formed a man's descendants into so many classes, that forced the men to pass at the proper time from one class into another, that established those rights of succession which existed between them? Were emancipated men still subject to their father's power? Or did men when emancipated come under some other power, which was to control their father as well as them? Sir Henry Maine has left these and other such questions in obscurity. All he tells us is that it seems to have been a *self-acting principle* that regulated the transference of men from class to class. Need it be said that nothing like this was ever found annexed to the Roman institution? Patria Potestas was exhausted with emancipation.

It would seem that Sir Henry Maine's theory is not consonant either with the facts or with Patria Potestas. And it leaves nearly every circumstance connected with this family organisation of the Irish in obscurity.

His second piece of evidence of Patria Potestas having existed among the Irish may be briefly disposed of.

In the "Cain" Law of Social Connexions* the question is discussed, How many kinds of Social Connexions are there among the Feini? The answer is that there are eight—the connexion of the chief with his "aigillne" tenants; that of the Church with her tenants of ecclesiastical lands; that of the father with

* *Ancient Laws of Ireland*, Vol. II., p. 343, *et seq.*

his daughter; that of a sister with her brother; that of a son with his mother; that of a foster-son with his foster-mother; that of a tutor with his pupils; and that of a man with a woman. The connexion of a father with his son is not one of the eight. It is stated that in each of these relationships one of the parties has over the other "judgment and proof and witness"—which the Editors of the *Senchus Mor* have amplified into the "*power of pronouncing* judgment and proof and witness"—under certain conditions as to credibility as regards women witnesses—over the other. In another passage * it is said that "the literary foster-father [or tutor] has *power of pronouncing* judgment and proof and witness upon the foster-pupil, as has the father upon his son, and the Church upon her tenant of ecclesiastical lands;" and but for this there would be nothing to show that a man had over his son the power, if power it was, which, as we have seen, his mother, his foster-mother, and his tutor (the latter a stranger to the family) alike possessed in relation to him.

Sir Henry Maine has seen in this passage a disclosure of Patria Potestas. But that his mother and other persons might possess this power, or whatever it was, over a man, as well as his father, is of itself enough to exclude this suggestion. Had this been a power possessed by fathers only, however, why must it have been Patria Potestas? Summarily to identify any and every power possessed by a father over his children, the nature and limits of which are unknown to us, with the

* *Ancient Laws of Ireland*, Vol. II., p. 349.

Roman institution can lead only to error and confusion. It is not always that the error can be detected so completely as it can be in this case.

For, if the passage first quoted leaves the meaning of the phrase "judgment and proof and witness" in doubt, we get further light as to its meaning from the Crith Gablach,* where, after the statement that there are seven divisions of social grades, the question is put, What is the division of social grades derived from? The answer is, that it is derived from the similitude of ecclesiastical orders, and that "good *corroborative* proof, or denial, or evidence, or judgment *is due* from each of them to the other." Drop out the word corroborative, which the translators have inserted, and we have here, about all the seven social grades, a statement identical with that upon which the inference of Patria Potestas has been founded. And it appears that the phrase "judgment and proof and witness" refers to nothing other than legal process.

It has been pointed out that the words "power of pronouncing," in the passages we have been noticing, do not belong to the text. They have been imported into it in an attempt to give it something like meaning. But the Editors of the *Senchus Mor* have not omitted to state that, whatever "judgment and proof and witness" might mean, they did not, in their opinion, mean Patria Potestas.† "The provisions of the Irish Family Law," they say, "do not appear to have any

* *Ancient Laws of Ireland,* Vol. IV., p. 299.

† *Ibid.,* Vol. II., p. iv. of preface.

connection with the ancient Roman Law. The Irish Law demands for the mother a position equal with the father, and there is no trace of the exercise of that arbitrary power which was wielded by a Roman father over the members of his family." This is a statement that there is in the ancient laws of Ireland not only no trace of Patria Potestas, but proof of its absence. Having that upon such authority, it seems unnecessary here to adduce the multitude of facts by which it can be supported. The student can easily test the statement for himself. He will find it very fully borne out.

CHAPTER X.

PATRIA POTESTAS—CONCLUDED.

BEFORE going further, it will be well to recapitulate the results arrived at in the five preceding chapters. And

1. Though Sir Henry Maine professes that his description of the Patriarchal Family is taken from the Scriptures, we have found that nothing like the family which he has described is to be discovered in the Scriptures, and that the history of the Hebrews is not to be reconciled with the Patriarchal Theory. After the Hebrews became a nation the Hebrew father certainly had not Patria Potestas, and the laws assured a place of high honour to the mother. Going further back, it appears certain that Jacob had not Patria Potestas over his family; and it can scarcely be doubted that the position to which he attained by service in his uncle Laban's household, was much the same as that of a beenah husband in Singalese or African villages. At no time do we find among the Hebrews, either in law or in practice, anything suggesting that a father had over all descended from him those despotic powers which Sir Henry Maine has so often recounted to us. To show the

absence of Patria Potestas was the utmost that immediately concerned us, and that has been done. But from the older Scriptures we have learned that to some effects—at any rate in some families of the Hebrew stock—the mother rather than the father seems to have been at one time the representative of the family; and that she certainly could hold property—since the bride's price was paid not to her husband but to her and her son. And we have found an explanation of such facts as this, and of such marriages as Jacob's, in the first Scriptural reference to marriage custom—which, if not misinterpreted by us, is by itself enough to prove that, so far as the Hebrews are concerned, the Patriarchal Theory is utterly mistaken. At any rate the Scriptures, while they appear to disclose to us more than one ancient form of the Family, certainly do not exhibit the Roman form.

2. We have found among the early Hindoos relations disclosed between the members of a family, which are at no single point consistent with the Patriarchal Theory. The father had certain powers over his household, no doubt; but these were defined by law, and not despotic but greatly limited, and he was answerable to the courts of justice for any abuse of them. The mother had in the family a position of the highest honour. Both she and her children could hold property independently of the father; and it can scarcely be doubted that, not only were adult sons co-owners with their father of the ancestral estate, but that, at any rate in some parts of India, they could force him to make

partition of it. In the Hindoo family as it appears to us through Manu, the father, though head of the household and custodian of its common property, and, though no doubt honoured and submitted to, was not a Paterfamilias. The family disclosed to us is of an ancient type, which is not the Roman type.

No doubt the oldest of Hindoo law-books deal with a social condition which is very far from primitive. The same may be said of law-books in general. But it is to these that Comparative Jurisprudence, as represented by Sir Henry Maine, commonly makes its appeal. Some reasons have been given for believing that, as regards India, an advocate of the Patriarchal Theory would only lose by carrying inquiry farther back.*

3. As regards the Slavs, there is so remarkable a dearth of information about their condition in early times that we have found ourselves reduced to thinking with bewilderment of the confident way in which Sir Henry Maine has professed to find in their "legal institutions" evidence for the Patriarchal Theory. We have found that such information as we have, while disclosing nothing as to the power of fathers among the early Slavs, gives not a little evidence against the Patriarchal Theory. We have found among the Slavs women in the place of Queen, of Regent, of Ambassador. We have found relationship on the mother's side acknowledged; and in the very first article of their earliest laws—which, however, are not very early, and perhaps not perfect evidence about them—we have seen, in the

* See chapters xvi. and xvii. on "Sonship among the Hindoos."

provision for the revenging of homicides, an acknowledgment of a sister's son as among the nearest kindred. There is, then, to say the least, no evidence that the family among the early Slavs was of the Roman type.

4. We have found that neither the Joint Family of India nor the House-Community of the South Slavonians has, within the times known to us, either sprung out of, or dissolved into, a family of the Roman type. In both, too, the organisation deemed necessary has been unfavourable to paternal authority, and has involved its being to a great degree superseded.

As to the Russian peasant household of recent times, which has often been regarded (without evidence) as giving something like the type of the ancient family among the Slavs, we have found that what was chiefly remarkable about it was the contrast between the powers of its head and the powers of a Paterfamilias. The head of such a household, even when he was the common progenitor, instead of being despotic master, was regarded as manager for all the adult members of the household; was actually called manager; and, in every transaction affecting property, was required to consult with his co-partners.

5. As to the indications of Patria Potestas which Sir Henry Maine believes he has discovered in Ireland, it can scarcely be deemed too much to say that they have been obtained through a most strange misapprehension of his authorities. His theory of the curious fourfold organisation of the family described in the Irish law-books neglects some, and is at variance with

others, of the few facts disclosed about that organisation. And it is as difficult to reconcile it with Patria Potestas as with the Irish texts. The phrase, constantly recurring in the law-books, which he has relied on as furnishing corroborative evidence, has been shown to be connected with legal process. But, were it not so, the fact that the mother and various other people had the power denoted by it, if a power, equally with the father, shows that it could not prove Patria Potestas, but would rather tend to prove the absence of it.

All the cases therefore in which Sir Henry Maine has averred, or seemed to aver, that direct evidence of Patria Potestas was forthcoming have been examined, and the result is that there is no evidence of Patria Potestas to be found in any one of them; and that, excepting among the Slavs (as to whom nothing perhaps can be said that is not liable to challenge), there is on the contrary in every case evidence of the absence of Patria Potestas. What can indirectly be established, as by the prevalence of Agnation, and an argument founded thereon, of course remains to be considered. But it is now plain that it is by means of indirect evidence, if at all, that the prevalence of Patria Potestas must be made out. So far as direct evidence goes, Patria Potestas appears to be of distinctively Roman growth.

"The proof of its former universality is incomplete so long as we consider it by itself," is what Sir Henry Maine has said of Patria Potestas. So far as legal evidence or historical evidence goes, the fact appears to be that this institution is absolutely unparalleled, and that there were at the utmost only three or four cases in which its existence could even be alleged. Sir Henry's statement is the more curious, seeing that he had not overlooked what Gaius has said of Patria Potestas.

What Gaius has said seems well worth pondering in an inquiry as to the diffusion of Patria Potestas. Here it is briefly, in Sir Henry Maine's words: "Gaius," he says, "describes the institution [Patria Potestas] as distinctively Roman. . . . . . Among the races understood to be comprised within the Roman Empire, Gaius could find none which exhibited an institution *resembling* the Roman 'Power of the Father,' except only the Asiatic Galatæ." *

Sir Henry Maine, be it said, does not question the evidence of Gaius. He has taken it as showing that the institution which, among the Romans, proved so durable was in its nature not durable, and so had *disappeared;* and thus has almost seemed to make out of its non-appearance anywhere a proof of its former universal prevalence.

But the opinion of Gaius was that Patria Potestas was distinctively Roman; and his opinion is the weightiest ancient opinion obtainable. He was one of the few Roman lawyers who were deeply interested in the antiquities of law, and given to tracing institutions to their origins. If there was any one living in the time of the Antonines likely to study carefully the laws and institutions of the numerous peoples whom, by that

* *Ancient Law,* pp. 135-6.

time, Roman conquest had brought within the scope of investigation, Gaius was the man. His statement, of course, involves that he had made a study of them. His opinion that Patria Potestas was distinctively Roman was formed upon the examination he had made of them. The validity of his evidence conceded, the reasons for his opinion are obvious, and they might well be deemed conclusive. With a vast area open to observation, containing races in very various stages of advancement, many of whom had been more or less under observation for centuries,* Gaius had found Patria Potestas nowhere, and had found only in one instance an institution which even resembled it. And Patria Potestas had in Rome proved a durable institution, and, indeed, singularly tenacious of life.

A modern writer has, in some cases, facts at his command not accessible to the ancients, which may exclude what, apart from them, was a just conclusion. But it has appeared that Sir Henry Maine is, equally with Gaius, without examples of Patria Potestas. And his views about it are only possible to one who deems such new facts as are available to modern writers unworthy of notice.

* There can be no doubt that Gaius was aware of what Cæsar (lib. 6, c. 19) had said of the Patria Potestas of the Gauls, and had rejected it as untrustworthy. Possibly he had more trustworthy information. But is it too much to pronounce it incredible that, with a system under which husbands and wives held their goods in common, the survivor of them having the right to the whole of the accumulated capital and interest, the husband should have the right to put the wife to death whenever he pleased?

To show the full weight of the statement of Gaius, let us recall what was the area open to his observation—that is, the extent of the Roman Empire under the Antonines and Marcus Aurelius. It included the whole of Southern and Western Europe (including Britain) as far as the Rhine and the Danube, with considerable territories beyond the latter river; Africa, to the Great Desert and the Cataracts of the Nile; Asia Minor and Syria as far as the Euphrates. And even beyond these limits there were countries, like Armenia and Mesopotamia, which the Romans knew well, and over which they exercised considerable authority and patronage.

That a writer like Gaius, whom there is no reason to suspect of carelessness in investigating facts, should have found no Patria Potestas within this vast range, is a fact most impressive when the question is raised whether Patria Potestas was a primeval and a universally prevalent institution.

Sir Henry Maine, while casting no doubt on the statement of Gaius, and not dwelling on it further than to point out to how great an extent Patria Potestas had *disappeared*, remarks that the diffusion of Roman citizenship among the races subject to Rome, which began with the Empire, and was completed under Antoninus Caracalla, "must have enormously enlarged the sphere of the Patria Potestas." He makes no attempt to show that such enlargement took place in fact—that the Roman institution did become naturalised among the various races contained within the Empire.

And, of course, that, had he shown it, could not have helped his main argument. His object was to exhibit the singular fortune of this institution—which, universal, *ex hypothesi*, at the beginning, but difficult to find because, *ex hypothesi*, not durable in its nature, was at last re-diffused from a people among whom it had proved singularly durable, and re-established—so far as an enactment designed to confer a privilege could establish it—throughout the Roman world. He does not seem to have noticed that he was raising up a crop of difficulties for himself in dealing with the races not at the time referred to within the Empire, had instances of Patria Potestas occurred among them. If so much diffusion were possible, could such instances be assumed to have been of native origin?

CHAPTER XI.

DERIVATIVE INSTITUTIONS. THE EVIDENCE AS TO THEIR ORIGIN WHICH MAY SUFFICE.

IN *Ancient Law* Sir Henry Maine relies for his proof of the Patriarchal Theory chiefly on Agnation, which, he declares, can be found (or traces of it) almost everywhere, and which, he says, "implies the former existence" of Patria Potestas wherever it is found. In a more recent production* he has somewhat qualified his original statement, according to which, wherever we find Agnation the paternal powers must have formerly existed. His later statement is that "where there is Agnation there must almost certainly have been paternal power." This change of terms shows a disposition to avoid appearing to overpress the point, but what change of view it involves is not altogether clear. It may mean that, at the date of the later writing, Sir Henry was not sure—but only almost sure—that there was between the two the nexus which he had supposed to exist between them ; or it may be an admission that perhaps Agnation could come into existence indepen-

* "South Slavonians and Rajpoots," *Nineteenth Century*, December, 1877. *Early Law and Custom*, p. 244.

dently of Patria Potestas. In case the latter be the true meaning, it will be well to point out that, except in the unqualified form, his proposition is of no value. What shall determine our judgment, if we find Agnation anywhere, as to whether it was preceded by Patria Potestas or not, if it was not in the nature of things necessarily preceded thereby?

But why should Agnation either imply, or almost certainly imply, the pre-existence of Patria Potestas? Certain reasons for this are assigned in *Ancient Law*, and these will be examined by-and-by. Let us now, however, consider, as briefly as may be—beginning with some of those cases which are simplest or easiest—on what grounds, on what kind and amount of evidence, the belief that one institution or practice is derived from another can reasonably be held with so much confidence that we may venture to say the one implies the pre-existence of the other.

The feu system of Scotland owes its existence to accident, the statute *Quia Emptores*, designed to prevent sub-infeudation, and which did prevent it in England, having become inoperative in the northern part of our island. It has flourished, and continues to flourish now that feudalism has virtually disappeared, on account of the permanency of the interests which it has established—the feus being perpetuities connected with the land, often carrying the right to costly buildings—and through its convenience, and especially the superiority of the feu as a building tenure over the lease. The name and the terms employed, as well as the nature of the tenure, still show the origin of the feu ; and though it may already be regarded as an independent institution, it is, in fact, an example of the feudal system in operation, which has been fixed and retained notwithstanding that the system has passed away.

It is probable that the feu system will long cling to the smaller perpetual holdings of land in Scotland. Were it then to last, just as it is, till memory of the feudal system having been established in Great Britain has passed away, might it be confidently recognised as an application of the feudal system, and the pre-existence of the feudal system be inferred from it? It might, on condition that full accounts of the feudal system remained; that it was known to have been rather widely diffused in Europe; and, à fortiori, if it was known to have been established among populations likely to have influenced inhabitants of Britain; and, of course (though in such a case this need scarcely be mentioned), that no preferable explanation was forthcoming. For, in the first place, the feu, in the case supposed, would not merely present some semblance of being a feudal holding—it would be feudal all over, feudal in every incident. The chance of there having been coincidence, the chance of so exact a resemblance in all of so many particulars (including highly technical terms) having been produced by other and unknown influences—the risk, that is, of mistake in accepting the suggestion of resemblance, which in such cases is what is to be feared, and what must be shown to be small—would be absolutely inappreciable.

Even if the known diffusion of feudalism left it improbable that feudalism had made its way into Britain, there would be, in the case put, no room for reasonable doubt as to the feudal origin of the feu. The less improbable the state of knowledge made that, however, the more confidently could this conclusion be accepted.

And, a feudal origin once admitted, it would not be difficult to judge between the two views between which choice would then lie. Either feudalism had been generally established, and a convenient application of it had been preserved; or a single application of it had been somehow, at some time, introduced. Probability decides in such matters, and probability would be immensely in favour of the former view. Again there might be confident conclusion. It is only when the intercourse of nations has become very free, and the development of intelligence leads to the comparison of institutions—and seldom then—that a people takes what suits it of a foreign system and leaves the rest. And what is likely to be taken in such a case is only the idea, enough to give the use which is desired; not the inconvenient details and the unfamiliar foreign terms. What we are always forced to reckon upon, therefore, is that people have applied the systems they have had and have grown familiar with. But, in the case supposed, apart from this consideration, the application would seem almost necessarily to imply the establishment of the feudal system. It would bear every appearance of being related to a system of which it was an application. And it would seem to the last degree unlikely that the correctness, to the minutest detail, which it would exhibit could be found without the system having been established and applications of it made familiar by daily use.

With such perfect resemblance, then, as has, in the case just put, been supposed between an existing institution or practice and the operation of an ancient system which has been fully described, and is known to have been considerably diffused in the same region in which the former is found, we could not hesitate to infer that the latter preceded the former in the particular place in which we find it. And we could not infer less than that the ancient system had been long enough and generally enough established in that place for it to have the power of impressing general practice, and so leaving an application of itself to future generations. It may be said, indeed, once for all that whenever we come upon a practice in general use which can be taken to be an application of an ancient system, less than this cannot be inferred as to the former prevalence of that system. In practice, we have in general —but by no means always—less perfect resemblance to guide us than there would be in the case supposed. That case shows, however, in a strong way—and it is chiefly for this reason it has been presented—the great value of resemblance, resemblance not in externals only, but in idea or purpose, methods, and effects, as a means of enabling us to connect a derivative with that from which it is derived. And with less than perfect resemblance—it being possible to exclude to a reasonable extent the chance of mistake—the indication of resemblance may often be distinct and strong enough to carry us to a highly probable conclusion.

On the other hand, there are numerous cases in which, the indication being not quite sufficient as regards either distinctness or quantity, unless corroborative evidence of some weight be obtainable, the notion that an institution or practice has been derived from another can only be at most a more or less amusing speculation.

Suppose, for example, as before, that at some future time it should be unknown that the feudal system was once established in Britain, and that there should then survive, not the feu system as it is now, but some remainder of it—the system, that is, docked more or less of the circumstances by which it could be identified, and perhaps in some respects altered. It would then depend upon the significance, and in some degree upon the amount, of those original characteristics which survived—somewhat, too, upon the extent to which methods or details had been modified or changed—whether or not there could be a hope of assigning the remainder with any positiveness to its origin, and then inferring the pre-existence of the feudal system in Britain. It might easily happen that there would be nothing left by which its feudal origin could be recognised. To give any chance of that being recognised there would have to be remaining something distinctively feudal, something for which no other origin could with equal probability be imagined. Even then there might not be enough to yield more than a surmise.

If, however, there remained enough of feudal characteristic to give distinct suggestion of a feudal origin, and there had been failure to account reasonably for this otherwise, it would forthwith be at least a plausible opinion—more or less plausible according to the circumstances—that what we had come upon was a remainder of feudalism. The more points there were at which it gave feudal suggestion, the less would be the likelihood of coincidence, the less the risk of mistake in accepting that suggestion; and the approach to a positive conclusion might be correspondingly nearer. (There are cases, however—but the case supposed could hardly be among them—in which evidence got at a single point may be absolutely convincing.) If now there could be found other practices or institutions which seemed also to give indication of feudal origin, the accumulation of these indications with those previously possessed might found a fairly probable inference that each and all of the institutions or practices dealt with had had a feudal origin. All this, of course, upon condition that full accounts of the feudal system remained, and that it was known to have been considerably diffused. If the state of knowledge made it probable, or left people free to think, that it had been established in Britain—if, for example, it had been established in a neighbouring country, or among a population likely to have influenced British institutions—an inference of fair probability might be arrived at the more readily. Where feudalism was known to have been established, such an inference could be got more easily still—less would suffice for it. And the argument in another case might be greatly helped by showing that, where feudalism had been established, there appeared indications of its influence like those the origin of which was being considered.

To pass now to a new case. The Common Law of Scotland did not allow its ordinary rule applicable to the property of spouses when one of them died and there were no children—which treated the personal property of both as, in virtue of the marriage, their joint property, and so divided it equally between the surviving spouse and the relatives of the deceased—to take effect unless either there had been a child born alive or the marriage had subsisted for a year and a day. Failing the subsistence of the marriage for a year and a day, and the birth of a child which lived (though it did not survive)—and, of course also, failing the birth of a child of the marriage after the husband's death—the law severed the interests which on marriage had become united; so that simply what the surviving spouse had brought into the common fund went to him or her, while what had been brought by the deceased went to his or her relatives. Now, in the old Celtic marriage by handfasting it was agreed that a man and woman should cohabit with a view to marriage for a year and a day; and they were husband and wife without further ceremony within that time if a child were born, or at the end of it, if the mother proved then to be with child. Otherwise, they were then free to leave one another and each to handfast with any other person; their interests, so far as they had become common, being of course severed in the event of their parting, and also in the event of one of them dying within the period of contract while there was no child.*

It appears then that in marriage, when one of the spouses died within what had been the handfasting term, the Scots Law applied its ordinary rule only when by custom handfasting would have turned into marriage; and that it did not apply it, but ordained a severance of interests—which would in the same circumstances have followed upon a handfasting—where handfasting would not have turned into marriage. Handfastings which outlived the year and day (which, except when one of the spouses died, must have commonly happened) were marriages; and within the year and day, in a certain event, handfastings became marriages. And in the cases which corresponded to these last, and in all cases in which marriage outlived the year and day the Scots Law applied its theory of marriage, which gave each spouse an interest in the property of the other; while it ordained that everything should be as if there had been no marriage in those cases in which there would have been no marriage with handfasting. Marriage within the year and day and Marriage not within the year and day were separate

* The accuracy of this statement of the contract may possibly be questioned, but, even if it be, that is immaterial as regards the use which is now to be made of it.

headings with the old text writers. And when within the year and day handfasting would not have become marriage, even when there had been a settlement, the ordinary rule for marriage did not apply—that is, the settlement was of no effect—except when there had been a stipulation to the contrary.

Upon what conditions, now, in the case stated, might we venture to conclude that the reason why marriage did not produce the full effects of marriage until it had lasted for a year and a day was that anciently, during that period, marriage was commonly inchoate? Had handfasting, as above described, been a prevalent practice, and been replaced by immediate marriage, the ordinary consequences of handfasting being nevertheless retained in the law, the law would have been just what it was. That is plain enough. And, on the other hand, no other explanation is known of those curious provisions; and it will be found extremely difficult to suggest one that will bear examination. Policy, indeed, might account for the year and day rule; but can it also account for the exception to that rule? Clearly not in a satisfactory way. And if it were known that the custom described commonly prevailed in connection with marriage among the early Scots, and the question were whether the provisions of the law were derived from it, there could be no reasonable doubt as to the answer.

Supposing that not known, however, how far would the coincidence between the law and the operation of the custom carry us? A custom of which accounts have come down to us prevailed of course to some extent. Might it be inferred forthwith, from the exact correspondence between the law and the operation of the custom, that the latter must have at any rate prevailed sufficiently to leave its mark upon the law? Most impressive as that correspondence is, so much could not be maintained. The custom gives us, no doubt, a perfect hypothesis to account for the law. There is identity between the provisions of the law and the operation of the custom. But there is in this case (so to speak) resemblance only in effects; and, perfect as it is, that cannot carry us so far as a resemblance which extends to idea and methods as well as to effects. Besides, coincidence has always to be reckoned with and guarded against. Before making such an inference, therefore, we should have to assure ourselves that the custom had had extension enough to make it not unlikely that it should affect the law of marriage. Having shown so much, we might trust to the evidence of resemblance for the rest. The custom being peculiar to the country where it occurred, more evidence would be needed as to its extension there than would be needed if it were more diffused. On the mere knowledge that there had been a custom which would have operated as has been described, however, it would be a very plausible opinion—an opinion clearly preferable to every other, and extremely likely to be confirmed by inquiry if evidence were obtainable—that the provisions of the law were derived from that custom, and therefore, that it had been pretty generally prevalent. And unless the evidence for the custom were wanting, not in volume only but in weight, it would be more. It is the weight of the evidence remaining as to the extension of the custom, which must determine the degree of plausibility or probability which would attach to such a conclusion. It might be held with something like confidence upon evidence a long way short of what would be needed to show that the custom had been generally established.

And, in considering evidence in such a matter, it would be unreasonable to dwell upon the absence of that which, from the nature of the case, could not be forthcoming—to shrink from a conclusion merely because only such evidence as happened to be recorded had come down. The weight of what remains is what would have to be looked at. Where the bulk of a people do not read or write, contracts will not often be made in writing; where there are no historians, customs will die out without having been described; and even when a rude people has its annalist, the customs of the people—familiar to him, though matter of curious inquiry to us—are about the last things he intentionally writes about. Nevertheless, we may have access to evidence which may be weighty and even conclusive. If a contract be in question, we may do pretty well if there have come down to us some of the written contracts made by the better sort of people—especially if it be a contract connected with marriage. We shall not do rashly in inferring that the contracts made between chiefs in connection with marriage had popular sanction—

and it is the contracts made by people of condition that are most likely to determine the subsequent law—or even in inferring that, though they may have varied from, they coincided in the main with common usage.

It is through contracts made between chiefly families, it may here be said, that we know of handfasting in the Highlands—where such contracts were made up to rather recent times; and if there be some risk of error in thinking that the chiefs did not invent this singular contract for their own exclusive use, and that they did not run counter to usage in entering into it, there seems to be no more than must be ventured whenever we form conclusions on probable reasoning. That a contract which, if not identical with, was barely distinguishable from, that of handfasting, was the common portal to marriage in some southern parts of Scotland—a grievous scandal in the eyes of the Reformed Clergy—till some time after the Reformation, tends strongly to support the view that handfasting, or something practically the same, was anciently popular usage; and would, of course, be a weighty fact in any argument on the connection between it and the Scots marriage law.

What we have now deduced is that where a custom found in a single country has been so fully described to us that we can understand how it must have operated, and we find later in that country, in connection with the same matter, exceptional arrangements, also peculiar to the country, identical in operation, it is at once a very plausible view—likely upon inquiry to prove preferable to any other—that the latter have been derived from the former; that this may be taken as made out upon evidence sufficient to show that the one had a certain chance of affecting the other, and, therefore, upon evidence a good deal short of what would be required to show that the custom was once generally prevalent; but that the degree of extension that can be made out for the custom must determine whether a conclusion to this effect may be made positively or not—whether, that is, we can have a reasonable conclusion or only a very plausible conjecture. In this case, nevertheless, as in the previous case, we see that, as a test of connection, perfect resemblance between the things compared is of very great value. In this case, supposing the resemblance less than perfect, even if we knew that the ancient custom had been generally established—though knowledge of that would carry us over trifling differences—there would always be room for question as to whether the provisions of the law were really derived from it.

Let us assume, now, the evidence for handfasting to be thought strong enough for a confident conclusion that the provisions of the Scots marriage law were a remainder of it. Could we then, were we to find marriage denied its effects in the same or similar circumstances in another country, venture to say with confidence that handfasting, or something similar, must have formerly prevailed there also? Could we say: there may have been handfasting where there is no such denial to marriage of its effects, but where there is such denial there must have been, or almost certainly has been, handfasting? Surely not. That view would be pleadable; it would be promising; one could not be sure but that it was right; we should look hopefully for means of verifying it; and even without these, many people might think it very plausible. But on belief in the connection of the two things founded on a single instance—the custom being known to us only in that instance—to be confident about it would be absurd. The custom would furnish a hypothesis for the new case than which none could be better. But even a perfect hypothesis cannot of itself carry us so far. And, on the other hand, the second case might be such as to shake our faith in the conclusion to which we had been brought in the first.

It would be different if it could be made probable that handfasting had left the remainder in a number of instances. On finding just the same thing in a new place—and no other explanation for it—then, with so peculiar a provision, it might, even on a small number of instances, be fairly probable that that, too, was a remainder of handfasting. If, in addition, it were known historically that handfasting had been a common custom, we might see our way to coming to a positive conclusion to that effect. Given, for example, handfasting known to have been so common that it would be no surprise to find it in any new place—or that, from our knowing it to have prevailed somewhere near, or among a kindred population, it would be no surprise to find it in a particular new place—in either case, on finding what in Scotland and some other places had been, for sufficient reason, taken to be a remainder of it, we might reasonably infer a former prevalence of handfasting. The more the cases in which it could be believed to have left the remainder, the more positive in the circumstances supposed would be our conclusion. But that conclusion might, in such circumstances, be reasonable, when the remainder had previously been connected with the custom only in a single instance.

Let us next—taking a case which has been matter of controversy—try to see on what conditions it may be legitimate to regard the Levirate as a relic of Tibetan Polyandry, from which it may be inferred that Tibetan Polyandry formerly prevailed wherever it is found.*

The Levirate (which we must be careful to distinguish from a very different and much less surprising thing—the custom whereby a brother succeeded to his elder brother's estate and to his wife as forming part of it) occurs only where monandry, separate property rights of individuals, and the succession of sons to fathers have been established—the last so firmly that even a fictitious son is preferred to a real brother—and where in general (in the great cases—the Hindoo and Hebrew—certainly) marriage with a deceased brother's wife is by rule forbidden. And the regular operation of it was to impose upon a man whose brother (in the

* See "The Levirate and Polyandry," by J. F. M'Lennan, and "A Short Rejoinder by Herbert Spencer," *Fortnightly Review*, 1877. See also chapter iv. of *Hindu Law and Usage*, by J. D. Mayne. Madras and London, 1878.

Jewish case, whose elder brother) had died, leaving a widow and no children, the obligation to take the widow and "raise up seed" to that brother. The son begotten by the surviving brother upon the widow was accounted his deceased brother's son, and as such—cutting out the Levir, his actual father—he succeeded to that brother's estate.* Adoption is an expedient a "Patriarchal Family" might have resorted to, as modern Hindoo families do, for the purpose in view, and it is an expedient not inconsistent with the conditions and moral sentiments proper to monandry. But what to ancient Hindoos and ancient Hebrews seemed right was, that the brother of the deceased should "perform the duty of an husband's brother unto" the widow—an expedient of a kind repugnant to monandry and shocking to the moral sentiments that grow round it.

Among the Hebrews the widow was at first her brother-in-law's wife without any ceremony, and she had a right to him as a husband; afterwards—no doubt under the influence of ideas of propriety derived from monandry—a formal marriage became indispensable. Among the Hindoos it was thought proper that the intercourse should come to an end when one or at most two sons had been born, and there never was a marriage.

* Among the Hindoos, and in early times (as the case of Judah and his daughter-in-law shows) among the Hebrews also, failing a brother, a near kinsman *could* act the part of a brother by the widow; but the Levirate has to be regarded as essentially an institution which imposed a duty upon brothers. The brother was the *proper* person to be Levir.

The Brahmin law-writers, though forced to tolerate the Levirate, because they found it an ancient and firmly rooted custom, loathed it as a practice "only fit for cattle;" and, just as the Jews imposed a marriage ceremony, they prescribe restrictions and regulations (we do not know to what extent they were generally submitted to) designed to mitigate the scandal of the practice — restrictions and regulations which, being avowedly designed for that purpose, could not, even had we not the Jewish case to enlighten us, be supposed to have accompanied it at first, when it can have caused no scandal. The feelings with which they regarded it— the feelings which a practice of monandry must produce —when they became general, put an end to it among both Hebrews and Hindoos.

We have here, then, along with monandry, a practice utterly antipathetic to it, which it ultimately destroyed, and which therefore it may safely be said it could not of itself have given birth to—the essential things in it, as ordinarily practised, being that, when a man died leaving a childless widow, it was his brother's duty to treat her as his wife, and that the brother's son by her was accounted the dead man's son in such good faith that he succeeded to the dead man's property. If it must have originated independently of monandry, it is in some custom which prevailed before monandry became a general practice that we shall do well to look for its origin. We are forced, or almost forced, to think that among Hindoos and Hebrews there were other and grosser marriage customs before monandry; and, once brought so far, there is no difficulty in thinking of the custom which could have supplied a basis for it. With Tibetan Polyandry, all the brothers of a family marry one woman; and the children of the brotherhood are regarded as children by all the brothers, while they are specially considered as the children of the eldest brother —perhaps in some degree because he may be the first to marry (his younger brothers becoming partners with him in marriage as they grow up as a matter of course), but chiefly because he is the head or administrator of the family.

Here we have a marriage system in which a man as he grows up becomes by custom husband of his brother's wife, and in which the children, by whomsoever begotten, are regarded as children by all the brothers, while they are specially the children of the eldest—and therefore a system which exhibits both the conception of marriage and the conception of fatherhood that appear with the Levirate. Repugnant to monandry, the Levirate, in its ordinary working, was, as far as it went, identical with Tibetan polyandry. Shocking to the feelings of propriety engendered by monandry, it exhibits without a shade of more or less the grossness of Tibetan polyandry. If we are forced to think that the Levirate had its origin in some prevalent marriage custom of grosser mould that flourished anterior to monandry, or along with it while it was yet not generally established, Tibetan polyandry stands out among marriage customs as the one which could have left to monandrous generations, which deemed it important that fathers should have sons to succeed them, such a remainder as it is.

Tibetan polyandry is known, too, to have been widely diffused throughout the world; and that a relic of it should long survive among a people who had practised it before and along with monandry cannot be thought surprising. Among a people who had practised it—and monandry along with it as occasion served—among whom monandry grew up alongside of it—the ideas and practices connected with it, even when monandry had become general, could scarcely disappear all at once. They would at any rate disappear in such a case—as polyandry itself would disappear—gradually and slowly. Would it be strange that, sons being desired for heirs, polyandrous practice should survive for a contingency in which the view of fatherhood associated with it could be turned to account, and that it should, monandry notwithstanding, receive the sanction of custom? If not, it is not surprising that Tibetan polyandry should leave behind it a remainder in the Levirate. Where it did so, from the nature of the case, there might remain little besides to attest its former prevalence.

In this case we have not, as in our initial supposition about the Scots feu system we had, an institution, otherwise extinct, surviving complete—just as it was originally—in one application of it. But what we have is not much less; for the same view of marriage, the same view of fatherhood and sonship—and these are all the elements in either—appears in both the

Levirate and Tibetan polyandry. And Tibetan polyandry, as has been said already, is known to have been widely diffused. In a polyandrous solution there is nothing surprising. There cannot, in this case, be such confident recognition that what we are trying to account for is an example of that to which we assign its origin as there might be in the former case. But there can be confident recognition of the substantial identity of the two. And, the Levirate being what it is, this is enough to carry us far. It becomes at once very much more than a plausible opinion that it was derived from Tibetan polyandry. For the extreme difficulty (it may almost be said the impossibility*) of suggesting for it a monandrous solution that will bear thinking about weighs heavily in favour of the polyandrous solution. Were it admitted that polyandry either preceded or went on side by side with monandry among either Hindoos or Hebrews, there could be no reasonable doubt that the Levirate was derived from it; and not only in the particular case. It being matter of common knowledge that polyandry has been considerably diffused, and the Levirate being the peculiar institution it is, one could not hesitate to believe that the one had been derived from the other in every case.

Since, however, the question is, On what evidence of the extension of the polyandry can we move on to the conclusion that the Levirate was derived from it? the first thing to be considered is what it is that has to

* See chapters xvi. and xvii. on "Sonship among the Hindoos."

be done? And what has to be done is to remove such doubt as there can be—and the difficulty of finding a monandrous solution taken into account, it cannot be of serious amount—that the resemblance, or rather the identity, perceived between the polyandry and the Levirate is only some strange effect of coincidence—produced by influences unknown to us, and which we cannot even imagine. That done, the Levirate itself will do the rest. For its evidence as to the former prevalence of polyandry—once we feel sure it can be relied upon—is conclusive. It has, at most, therefore, to be made probable that Tibetan polyandry has had a certain chance of influencing marriage custom where the Levirate has been found. To show, from the known diffusion of polyandry, that it is not unlikely to have preceded where the Levirate is found might almost be enough: to show, for some leading example of the Levirate, that there is some good evidence that polyandry preceded—to show, for example, that there are striking instances recorded which prove that it was practised in early times, and, being then unusual, was then justified as ancient custom; and that there are, besides, laws or usages (and one might suffice) which seem to be, as much as the Levirate itself, of polyandrous origin—would, it would seem, be amply enough, enough to justify a decided conclusion.

As has been pointed out in considering another matter, it is idle in such a case to dwell upon the absence of evidence which, from the nature of the case, could not be forthcoming. And it is the weight of such evidence as there is that is to be looked to, not the quantity of it. If we find that even one law or usage has come down which seems to be, as clearly as the Levirate itself, of polyandrous origin, and as difficult to account for otherwise, it is plain that the chance of coincidence, which is what we have to be on our guard against—in this matter insignificant at best—is so materially diminished that it may be disregarded. With a very few such it would become infinitesimal. Add to even one such law or usage good evidence, even if limited in quantity, of polyandry in ancient times and then justified as ancient custom, and we may without hesitation discard the fear of coincidence—and hold that the Levirate was derived from polyandry, and that the former prevalence of polyandry may be inferred from it.

Of course, the greater the quantity of evidence the better. But it is a surprising thing that there is any—for what there is has come to us almost by miracle. Brahmin poets and lawyers and their editors would have suppressed it all had they been able. And what we chiefly have to consider is how much room there is for mistake as to the indication given by such evidence as there is.

It need scarcely be said that such an argument as has just been described would gain in conclusiveness—though possibly not so much as it might gain by being materially strengthened in one case—so far as it could be repeated, for other peoples which have had the Levirate. The evidence from all the cases could be accumulated to dispel doubt as to the origin of the Levirate. Every custom, too, which shows a "note" of the polyandry, or is akin to it, or which seems to have come out of the layer of social condition to which the polyandry can be assigned, which can be cited from peoples who have had the Levirate, would add to the strength of such an argument. And evidence pointing to the polyandrous origin of the Levirate that would tell strongly, that might even by itself be enough for belief, might legitimately be brought from races that are backward beyond what either Hindoos or Hebrews were when history makes them known to us.

As the Patriarchal Theory is the subject of this work, there can be no harm in adding that the Patriarchal Theory has the Levirate to reckon with—and that, unless it can account for the Levirate consistently with "Patriarchal" custom, it is at once out of the field.

The cases which have, up to this, been examined have been all of the class that is easiest to deal with—though neither of the simplest, nor of the most difficult, order of cases of that class. They have been of the class of cases in which the connection of the derivative with that from which it is derived is suggested, and more or less attested, by resemblances which can be perceived between them—resemblances in idea, purpose, methods, results, and outward form. And, we have found that, for confidence in these cases—such confidence as Sir Henry Maine has in the connection between Patria Potestas and Agnation—there is needed a weight of corroborative evidence which is not usually forthcoming.

When, indeed, the older institution is known to have preceded in the place where the supposed derivative is found, and the resemblance is tolerably complete, a conclusion as to the connection between them is not very difficult to arrive at. But much must depend upon the degree of resemblance. If, on the other hand, that is not known, or not likely to be generally admitted, even when there is no deficiency of resemblance, when even the later institution or usage is substantially identical with the older, a good deal—varying according to the case and the chance it affords to other explanations—has to be done by way of corroboration, and to exclude possibilities of mistake. It must be known, first of all, that the older has been so far diffused that it is not unreasonable to seek in it an explanation of the later. It has then, at least, to be shown that it has been so diffused that it is not improbable it should have had its day where the later is found—and upon that one could not in many cases feel *confidence*. Unless, in some of the places where the later institution or usage is found, we find either direct evidence of the prevalence of the older, or other unmistakable relics of it—such as, by themselves, are good evidence for proving its former prevalence—in general there cannot be room for anything approaching to confidence. The methods by which an argument to establish the connection between the two may be strengthened, need not again be dwelt upon.

The cases of the class we have been dealing with which are the least difficult are happily those which, for throwing light upon the history of mankind, are of all the class the most important. They are those in which that which has to be accounted for is visibly a mimicry or symbol. In other cases of this class it can be sometimes perceived, and more often surmised, that convenience has at least favoured the preservation of some part, or of some use, of an old institution. But, owing to our incapacity—whatever may be its cause—to drop at once a mode of proceeding which has long been customary, practices which have been long common or general practices where they have prevailed, but have been by degrees superseded as needless or ineffective, often survive in a symbolism—which itself is usually inconvenient and ridiculous. That can in numerous cases be shown on direct evidence, and can be illustrated from nearly every field of human activity. For our present purpose it may be taken as admitted and beyond doubt.

When we come upon such symbolism in good preservation—not too far decayed to tell its tale plainly—when we find what is visibly and beyond mistake a mimicry or counterfeiting, something feigned to be done, not to be taken as done in earnest, and see that it counterfeits something that mankind have been wont to do in earnest, it is but seldom we need delay much before accepting the indication given that there is now a counterfeiting because there was once in the same place a reality of the same sort—that what we find people feigning to do, predecessors of theirs did in earnest. Where symbolism clear in its indication is found, there is scarcely ever found any customary explanation of it (though there is sometimes a half-consciousness of the true one): what is done people do unthinkingly, in obedience to usage, because it was done by their forefathers. And such explanation as is offered by speculative persons in ignorance of the tendency to symbolism is only rarely even stateable; so that in general the explanation from symbolism has such advantage as can be derived from the absence of any other fit to compete with it. It is not only sufficient, and, so to speak, visibly sufficient, and based upon a well-known tendency of mankind, but it has the field to itself. Except as being a mimicry of something that has become obsolete, what happens is unexplained, if not inexplicable. It is evidently a mimicry. And it being a way of mankind often to preserve in mimicry their discarded usages, it may in general, as has been said, be readily concluded that it is a mimicry of discarded usage. A certain diffusion among mankind at some time—or among people who have the symbol—of that which is counterfeited must, of course, be shown; and also that where that occurred, it was common practice—so common that its leaving its impress upon usage need not be deemed surprising. As to the amount of evidence to these effects that may suffice for a positive conclusion, that must vary with the case.

Where, instead of a complete symbol, we have only what seems the remains of a symbol—where the indication given is faint instead of clear—as may happen in the decay of a symbolism, of course there cannot be anything like confidence felt in a conclusion reached upon that alone. But, if there be clear cases of a symbol as well as faint—if it be known, not only that the obsolete usage has been widely diffused, but that it has frequently been symbolised—the previous prevalence of the usage among a particular people, may be inferred with much probability even from a symbol much decayed. When, on the other hand, we meet with cases, as may occasionally be done, in which the old usage, or a remainder of it, and its symbol are found side by side, and the one, as it were, passing into the other—when, for example, in connection with marriage, either a frequent practice of actual capture or of abduction without leave (which is a modification of actual capture) is found side by side with abduction by arrangement,* that is, the form or symbol of capture—these may be, even by themselves, convincing, yielding as good evidence as could be demanded for any conclusion.

What are, on the whole, the most difficult matters of the class we have been considering, are also, as it happens, on the whole the least important.

We have dealt mainly with cases in which, of an institution or usage which, after gradually shrinking, has given place to something else, custom has retained some part or some use—what remains preserving more or less recognisably the semblance of that from which

* See pp. 76 and 298–9.

it came. When we turn from these to cases in which we are disposed to think that an old institution (instead of shrinking) has been developed, or rather improved and altered, so as to make it serve new uses of men—when the problem is, let us say, to make out the connection between an existing institution and something which looks as if it might have been its germ—while the conditions of proof remain precisely the same, on the whole, much more care is needed to exclude possibilities of mistake—to exclude, that is, other explanations. A real resemblance, indeed—a resemblance, not in outward look only, but in idea and purpose—will still be very helpful. The resemblances that are found, however, are commonly vague and faint. And, with a real resemblance, we have to be greatly on our guard—so strong, on the one hand, in the cases of this order that we commonly have to do with (say, modern cases), are the chances against so much of resemblance having endured; and so many, on the other hand, the possibilities of the resemblance having come to the later institution from some other source. Custom works slowly and works gently—changing only where there must be change, sparing whatever can be spared, regardless of congruousness of parts or symmetry of form. But when men in conscious activity have, through official agencies like courts of law or parliaments, been transforming an institution for ages—applying it to new uses, discarding it for old uses, changing some of its methods, replacing others by more efficient ones—though congruity and symmetry may not have been much thought of, and there may sometimes be left a point or two which can be with some confidence identified as belonging to the original, it is indeed surprising if, on the whole, there remains a real resemblance to the original at the end. And, on the other hand, before institutions are much developed, societies have grown rather complex. Races, each starting with its own customs, have perhaps been mixed together. There has been, at any rate, intercourse with foreign peoples, and more or less imitation of foreign ways. Above all, where the resources of older civilisations could be drawn upon they will have been drawn upon, and only minute investigation can disclose to what extent the borrowing has gone. No doubt, it is possible to guard against all those sources of mistake, and sometimes that is done. And this order of cases has its favourable side. Where they arise, much evidence as to the origin of an institution is often, with careful and skilled investigation, obtainable.

A fair example of them, and an illustration of what has been said of them, may be found in our system of trial by jury and its origin. Trial by jury was long customarily assigned, on the evidence of some external resemblances, to an old English germ. It seems now to be settled that it has come to us through a form of process which came over with the Normans, and which came to them somehow from the Civil Law.

Where resemblance to another institution is altogether wanting, however, the difficulty of finding an origin for an institution is, in general, much greater than it is in the worst cases in which some resemblance is present. In these we have at least a clue to guide our inquiries—a hypothesis ready to our hands. There may be risk of our accepting it too easily; but, in any inquiry, it is of great assistance to start with a hypothesis. Where, on the other hand, the indication of resemblance is wanting, unless, by taking trouble, we can so far make out the history of the institution as to find in it something to help us, we must be content to be ignorant—or to make surmises which, however amusing, can lay no claim to probability. We may find, however, something, in a prior form of that which we are inquiring about, which gives us the clue of resemblance, which found, the rest of its history may be more easily and more hopefully followed up. Or we may find it in its earlier history, where we have been tracing it, so connected with another institution that we are put on the alert to discover whether that other had to do with its first appearance, and what it had to do with it. The question arises: what influence coming from the one institution, if any, could have originated the other? and, at the same time, this other question: does the institution which we are disposed to think secondary or derivative first appear along with the other in all cases in which we find it; or does it sometimes appear independently of that other, or in connection with something else? It is natural to look at the former question first, since it may give us a hypothesis; while the latter puts us upon fresh inquiry. And it may turn out that no plausible theory as to the connection of the institutions is to be had; or that there are half-a-dozen that are equally plausible, which is much the same thing; or, on the other hand, that a theory can be framed which looks so sufficient that it may be taken provisionally as a hypothesis.

If this last should happen, the theory may connect the two institutions either through something that is essential to, or distinctive of, the older, or through something which, though found with it, belongs to it only accidentally, and which might happen to occur apart from it. In the latter case, it may be possible to find the derivative apart from the institution in connection with which its origin has been inquired into; and such verification as the hypothesis admits of, consists in so finding it, and then finding along with it the accidental element in that institution with which the hypothesis connects it. It need scarcely be said that we can never get, in such a case, anything like, or equivalent to, "the power of predicting," except for that which is truly the cause; and that the occurrence of the derived institution in a new region, without the institution to which we first traced it, cannot raise any presumption that the latter also prevailed in that region. It may make it worth our while to look for traces of this institution. It may raise some hope of finding them. But that is all. The conditions affecting the former case, as the question which is presently to be considered comes under them, may, at the risk of being tedious, be set forth with more minuteness.

When we deduce the secondary institution, then, out of something essential or peculiar to the institution to which it has been traced, the question arises whether it has always first appeared along with this institution. It is by showing this that the hypothesis may be verified. And, however plausible it may be, unless verification can be extended to a good many cases, one cannot reasonably have any feeling like confidence in the hypothesis. The chances of having to give up our confidence are too great for us to be justified in feeling any confidence. For it may turn out that a totally different account can be given of the institution which, even as a hypothesis, may be preferable. And a single clearly hostile fact upsets a hypothesis. We know that hypotheses which have been thought beautifully complete have, in countless numbers, been set aside for others which, as soon as they had been stated, were perceived to be better; or have been discarded as clearly bad because of some fact which had not been known or had been overlooked. The position of a hypothesis which is unverifiable—the means of testing which are not to be had—if there be such—is, of course, very bad. For while there may appear reason for giving it up altogether, it can never possibly get beyond being a more or less plausible speculation. And if circumstances only permit of a hypothesis being tested in a very few cases, we must be content with claiming for it a corresponding degree of probability.

Of facts which are to be considered hostile only one or two need be mentioned. If we find the institution we are inquiring about existing independently of that from which our hypothesis derives it, and in connection with something else, that is *primâ facie* a hostile fact. It may be possible to explain this; but, unless that be done, it is a fact which stops the way against the hypothesis. For the explanation a subsidiary hypothesis is required; and that, too, has to be tested and verified. It is indispensable, that is, to show that, in a number of cases in which the two institutions have been found together, the derived has outlived the parent institution —and outlived it apparently for the reason suggested. This sufficiently done, what was *primâ facie* a hostile fact may be turned into a friendly one. But to find the supposed derivative in independence of that from which a hypothesis deduces it must always, in the first instance, raise doubt as to the goodness of the hypothesis. If it is often so found, there is much probability that we are altogether on the wrong track—that we have been seeking the origin of our institution in the wrong place—that the appearance of special connection between the two institutions to which we have given ourselves up has been delusive.

Again, if instances of the supposed derivative are found to be much more numerous than instances of that which we take to be the parent institution, that also is *primâ facie* a hostile fact. The latter institution, when fully developed, should perhaps always, unless interfered with, have produced the former; but, as it may sometimes have failed of that, it should have occurred, if anything, the more frequently of the two. Where, then, it is found more rarely, explanation is necessary. It may be possible to give it. But here again the probability arises that we have got on a wrong track.

Only one more example of hostile facts need be given. If that which we take to be the originating institution is ever found without the supposed derivative, obviously that has to be reckoned with. It will supply a ready test of the goodness of our hypothesis. In all probability it will dispose of it altogether.

Seeing that so much has to be done, it must be evident that, in inquiries in which evidence is not often to be had in abundance, there can be confident conclusion in the end only in rare cases. And, instead of having in the supposed derivative, when found apart from the institution from which we are deducing it, an easy means of proving the former prevalence of that institution, it will be fortunate for us if we are able to show that its being so found apart does not force us to give up our hypothesis.

All this is elementary—and has been intentionally pitched low. It will be seen by-and-by that it has not been set down without necessity.

Of an institution associated with another through an accidental nexus, a sufficiently good illustration may be found in primogeniture. The origin of primogeniture in Europe has been, upon a certain amount of evidence (the sufficiency of which is unimportant to us here), traced back to the feudal system. But if the feudal system did establish primogeniture in Europe, this happened through those who moulded that system having resolved that fiefs should be impartible, and held by one person, and through it appearing reasonable or convenient that what one only was to have should go to the eldest. The impartibility of fiefs was not of the essence of feudalism—and still less that there should be a preference of the eldest; and therefore, on this view, it was by accident that primogeniture arose in connection with feudalism. If this account of the origin of primogeniture in Europe be a true one, primogeniture might arise wherever a feeling prevailed that some subject of inheritance ought not to be divided (or where dividing it had never been thought of); and the truth of the account is to be verified (so far as anything can be done for it by verification) by showing that succession to things regarded as not divisible has been given to one son in other cases, apart from feudalism.

Verification of the hypothesis, however, though necessary, cannot carry us very far in such a case. To be sure that the hypothesis is applicable to the case is what is most important, and verification of it cannot help in this. For primogeniture in Europe may have been a result of impartibility, and may have been found with the feudal system, and yet may not have first arisen in Europe in connection with the feudal system. It may have arisen earlier. In such a case, a high degree of probability is always likely to be unattainable.

In cases of this class, it will usually be found that the secondary institution is of a different nature from that in connection with which it originated; is without resemblance to it; and has its own separate objects and effects. Special connection between the two is not readily suspected. It has always to be made out pretty closely upon evidence. It need scarcely be added that, from finding primogeniture in Africa, we should get no indication of feudalism having prevailed there. It is impartibility that would be indicated.

Of an institution connected with another, not by an accidental nexus, but essentially, we have, according to Sir Henry Maine, an exceedingly strong example in Agnation. We need not think, then, of other examples. Agnation, on Sir Henry's view, proceeded from Patria Potestas; arising with it necessarily, that is, always; determined by its very nature, and being, therefore, it might be thought, dependent thereupon. Patria Potestas, on that view, always produced Agnation, and nothing else ever produced it—the one necessarily or invariably issued from and accompanied the other. Now, to establish a connection of this kind, what may suffice? Speaking generally, it may be said, as strong reasons as are needed for any proposition whatever. We have already seen what are the particular questions which arise. Does the hypothesis seem a sufficient one? Is it the best hypothesis that is to be had? And it should be, in this case, not only the best, but beyond all doubt the best. Then, has it been verified in many cases? And, in this case, an unusual amount of verification is clearly indispensable. Lastly, is there an absence of hostile facts; or have facts which were *primâ facie* hostile been sufficiently accounted for? If all these questions could be answered satisfactorily, Sir Henry Maine would undoubtedly have reason for confidence. But confidence as to what? If we knew nothing of his case we should expect him to be confident that Patria Potestas and Agnation were always to be found together—that wherever he found the one he would also find the other.

But the question we had to put at starting was, why should Agnation imply the pre-existence of Patria Potestas? Sir Henry Maine thinks that Agnation can often be found in independence of Patria Potestas, where Patria Potestas is not to be found or even to be traced, and in connection with other forms of the paternal power. What he is confident about is that, where he so finds it, he can infer the pre-existence of Patria Potestas. And it is only as a means of recovering lost cases of Patria Potestas that its connection with Patria Potestas is of importance to him. That Agnation should be found without Patria Potestas, and in connection with other forms of the paternal power, however, and especially that it should be so found "almost everywhere," is *primâ facie* a fact hostile to his hypothesis—and, the nature of the hypothesis considered, a fact extremely hostile. *Primâ facie*, this raises a strong presumption that the explanation of

Agnation is to be looked for in some other direction. No doubt such seemingly hostile facts can sometimes be accounted for and made friendly. The difficulty of doing this with Agnation seems unusually great. But we have seen how it is to be attempted. There is needed a new theory or hypothesis to explain how Agnation survived that which gave it being, and on which the main hypothesis made it seem absolutely dependent. And, at the least, it is indispensable to show that in some cases in which Patria Potestas and Agnation have been found together, Agnation survived Patria Potestas, kept up by the influence indicated in this new hypothesis. It should appear, too, that this influence was not likely to have acted in exceptional cases only. With a second theory so verified, there might, after all, be more or less confident inference of a pre-existence of Patria Potestas from Agnation.

It may be well to point out here, with reference to Sir Henry Maine's hypothesis, that Patria Potestas and Agnation dealt with entirely different matters, and did not act within the same limits. The one embodied the powers of the father. The other defined relationship, and through that determined rights of inheritance. Patria Potestas acted only within the family—was confined to the descendants of a single living Paterfamilias. Agnation had a far wider sphere; it connected the members of a family with one another, but it also connected the family with many other families. It might be not unreasonable to suppose that the father's power was placed so high as it was at Rome simply in the interest of fathers; and, in fact, Patria Potestas was extended to plebeian fathers at Rome as a privilege. Whether it could pass its bounds and produce effects far beyond them—effects, too, with which the degree of a father's powers has no obvious connection—is one of the questions we shall immediately have to consider.

CHAPTER XII.

PATRIA POTESTAS AND AGNATION.

In considering the hypothesis which derives Patria Potestas from Agnation it will be convenient to invert the natural order, and notice first what can be gone through most quickly and easily—the facts which bear upon the hypothesis. The hypothesis, as has been said already, involves that Agnation necessarily or invariably proceeded from and accompanied Patria Potestas. And, nevertheless, its importance to Sir Henry Maine entirely arises out of his thinking that Agnation, or traces of it, can be found almost everywhere, in independence of Patria Potestas, and in connection with lower forms of the paternal power. He says that its being so found "implies the former existence" of Patria Potestas.

To begin, then, Patria Potestas and Agnation are found together only in a single case. They occurred together at Rome, and have never been shown to have occurred together anywhere else. And to prove that the one always arose out of, and was thereafter to be found along with, the other, there is the fact that we find them together in one case.

To prove what it is equally necessary to prove, however—that Agnation, so originated, could outlast Patria Potestas—there is no fact at all. And the only fact that bears upon the matter tends, and tends decidedly, the other way. Sir Henry Maine supposes that Agnation, though originated as he thinks it was, could outlive Patria Potestas; and that it constantly did survive Patria Potestas for so long a period that every trace of Patria Potestas having occurred along with it had time to disappear. And yet at Rome, where alone he can show us the two together, it was Patria Potestas that showed the greater power of lasting. The last vestiges of the two disappeared from the law together. But, in fact, Agnation went first. The paternal powers were susceptible of abridgment and restriction in various ways short of extinction. The wife might become free from them; the children also; and yet they might remain for the slaves. And it was thus gradually that they perished. But Agnation is perfect, or it ceases to be Agnation. And the moment the ties of blood through women received civil effects Agnation was no more. This took place under the Prætors, long before the disappearance of Patria Potestas. As was to be expected, too, considering the difference in kind of the matters with which each had to do, the enactments which destroyed the one were mostly unrelated to those which broke down the other. Patria Potestas at Rome (where alone we know it) proved, in fact, an extremely durable institution. And if a judgment may be formed upon the one case which affords us the means of judging, it was everywhere more likely to outlast Agnation than Agnation to outlast it. At any rate, as the matter stands, there seems a lack of reason for believing—with confidence—that Agnation, if produced as it is said to have been, would generally be the survivor, and that it would survive Patria Potestas for ages.

We have now got the length of seeing that the hypothesis which derives Agnation from Patria Potestas is a naked hypothesis; one, too, from its nature, needing a great deal of verification and support, while it has not in the slightest degree been verified or supported. And, if cases of Agnation found with lower forms of the paternal power than Patria Potestas were really plentiful—their being so found being unaccounted for—it might be dismissed forthwith as manifestly a bad hypothesis. The facts would be that Agnation occurred, indifferently, so far as could be seen, along with various degrees of the paternal power; that it occurred with the exceptional degree of paternal power called Patria Potestas in the one case in which it has been found; and that, in this case, it did not survive Patria Potestas. For a hypothesis which derived it, on the strength of this one case, from Patria Potestas, and accounted for the numerous cases where it occurred, in which the paternal power was lower, by attributing to it a longevity indefinitely greater than that of Patria Potestas, nothing could be said. For a reason that will appear hereafter, however, this will not be insisted upon here.

A hypothesis that cannot be verified or supported can never be more than a mere speculation, upon which nothing can reasonably be founded; but it may be plausible or the reverse. Let us now see which of the two this hypothesis is.

Sir Henry Maine's argument on the connection between Patria Potestas and Agnation, as given in *Ancient Law*, is contained in the following passages:

(*a*) "The foundation of Agnation is not the marriage of father and mother, but the authority of the father. All persons are Agnatically connected who are under the same paternal power, or who have been under it, or who might have been under it if their lineal ancestor had lived long enough to exercise his empire. In truth, in the primitive view, Relationship is exactly limited by Patria Potestas. Where the Potestas begins, kinship begins, and, therefore, adoptive relatives are among the kindred. Where the Potestas ends, kinship ends; so that a son emancipated by his father loses all rights of Agnation. And here we have the reason why the descendants of females are outside the limits of archaic kinship. If a woman died unmarried, she could have no legitimate descendants. If she married, her children fell under the Patria Potestas, not of her father, but of her husband, and thus were lost to her own family." Then follows the well-known passage to the effect that early societies would obviously have been confounded if men had called themselves relatives of their mothers' relatives.—*Ancient Law*, p. 149.

(*b*) "The Parental Powers proper are extinguished by the death of the parent, but Agnation is, as it were, a mould which retains their imprint after they have ceased to exist. . . . . The Powers themselves are discernible in comparatively few monuments of ancient law, but Agnatic Relationship, which implies their former existence, is discoverable almost everywhere."—*Ancient Law*, p. 150.

Those passages, like many other explanatory passages in Sir Henry Maine's writings, do not yield a meaning that can be stated in plain terms.

That we are to understand from them that Patria Potestas, out of its very nature, determined relationship, and that the relationship it determined was always Agnation is, however, clear enough. And it is also clear that Sir Henry says it was subjection to the same paternal power, that made the first relationship.

How, then, was this primitive relationship expanded into Agnation as we know it? Patria Potestas, at its widest stretch, could cover but a small part of the limits of Agnation—for Agnation extended, roughly speaking, to all persons descended through males from the same male ancestor. The relationship limited by Patria Potestas must, at the utmost, have been very far short of the extent of Agnation. It is in vain we try to make out from *Ancient Law* how Sir Henry Maine connected those two. He tells us, indeed, that "Agnation is, as it were, a mould" which retains the imprint of the paternal powers after they have ceased to exist. But figures of that kind are nearly always delusive; and it is easy to see that this one is. It is inapt even for suggesting what Agnation was—relationship subsisting between descendants through males of the same male ancestor. But, at best, it only points to the fact that the relationship which, at Rome, connected families, and the persons included in them, with one another, was transmitted through the male members of the families. Whatever the origin of Agnation may have been, how otherwise could it be transmitted? That relationship was transmitted from father to children is, however, literally all there is to say for the view that the Power of the Father, acting as it were through a mould, was the determinant of Roman relationship. Sir Henry Maine should have at least shown us here how Patria Potestas acquired the capacity of leaving an imprint. For he assumes that it had not this capacity at first. He tells us that at first relationship began and ended with Patria Potestas.

Here occurs the question whether his primitive relationship can, even by a strain of language, be termed relationship. It was constituted by subjection to the same paternal power. It began and ended with Patria Potestas. A son therefore lost it when he was emancipated, or when he emancipated himself. A daughter lost it when she married, because she then passed out of her father's Potestas. Her children, we are told, never had relationship to her family because they were never under that Potestas. (But, when a Paterfamilias had "children, grandchildren, and great-grandchildren," might not his female descendants, or many of them, have married, and their children been born *within* the Potestas? Did not the Patriarch Abraham marry his sister-german?) And further, when the Paterfamilias died, his Potestas being at an end, the relationship which it had made ended too. Those who had been members of his family ceased to be related to each other. In its time of primitive vigour, that is, Patria Potestas left no imprint behind it. All the relationship then known was the connection made between people by their living under the same family chief. And that lasted only as long as they lived under him. The connection between a ship's crew when at sea under the same captain, or that of a regiment under the same colonel, has as good a claim to be called relationship as the connection between the members of Sir Henry Maine's primitive family.*

It was, therefore, something that seems a good deal short of true relationship that Sir Henry Maine had to connect with Agnation. He leaves us in *Ancient Law* to puzzle out for ourselves how the one could grow into the other. And to guide us, there is nothing except that (on his own assumption) the primitive relationship was not "as it were, a mould" which retained the imprint of paternal powers which had become extinct, but a passing connection, which might

* What Sir Henry Maine says of the emancipated son is that he lost "all his rights of Agnation." But this cannot mean less than that he lost relationship altogether. For otherwise relationship and Patria Potestas were not commensurate—relationship did not end where Potestas ended—in the primitive family. And (though it is not stated) the relationship of the family must have been at an end when the Paterfamilias died. For otherwise relationship did not begin and end with Patria Potestas in the primitive family—even the primitive relationship extended beyond the primitive family.

be broken in various ways, and which, on the death of the Paterfamilias, came completely to an end, leaving no trace of itself behind. Did Patria Potestas acquire in time a capacity which it had not at first? Or did something help it—and was it in virtue of that help that Agnation was produced? Or did relationship grow up in time after its own ways, independently of Patria Potestas, but somehow determined by it? Of these and other such views that may occur to us *Ancient Law* leaves us to take our choice.*

Of course, it is the exclusions of Agnation that

* A passage in Sir Henry Maine's latest work, *Early Law and Custom*, which has already been quoted (*supra*, page 22), appears to show that Sir Henry is now of the opinion—the only rational opinion, though not embodied in the account of relationship given in *Ancient Law*—that reflection upon the facts of blood-connection introduced systems of relationship. Casually, while gently blaming Mr. Lewis H. Morgan for over-facility of assertion and for a lax (though convenient) use of the term gens, he reminds him that "what was new at a certain stage of the history of all or a portion of the human race [apparently, the recognition of relationship through the father] must have occurred, not in connection with the gens, but in connection with the family;" and adds that, "at some point of time, some change of surrounding circumstances enabled paternity, which had always existed, to be mentally contemplated, and further, as a consequence of its recognition, enabled the kinship flowing from common paternity to be mentally contemplated also." That "when paternity reappeared," the kinship recognised was kinship through male descents only, is still, in this passage, ascribed to Patria Potestas (the reappearance of paternity "in association with Power and Protection"); but its *modus operandi* is not explained.

The passage gives, or rather suggests to us, a perfectly good reason for the acknowledgment of relationship between persons descended from the same ancestor. Apart from this, the view it gives is that relationship began in the family, and was, through the influence of chiefly need to be accounted for. And it is when we come to consider what is said of these that the difficulties in the way of the solution suggested in *Ancient Law*, and the need in connection with them of explanations that are not given, can be most plainly seen. That persons actually of the same family, under the same father, should be each other's relatives is not surprising, whatever our theory of the origin of relationship; and that the children of such persons should be each other's relatives would not be surprising. That which is remarkable in connection with Agnation, that which any theory of Agnation ought clearly to account

Patria Potestas, traced through male descents only—the view of *Ancient Law*, though without the explanation which accompanies it, and without alternative explanations. It leaves us, therefore, to go back to *Ancient Law* for explanation of what is the really important matter—the limitation on relationship ascribed to Patria Potestas—the exclusion of a man's descendants through females from relationship with his family. Everywhere else, it may be said, throughout Sir Henry Maine's writings of date subsequent to *Ancient Law*, the connection which he alleges between Patria Potestas and Agnation is taken as established, without there being anything to establish it except the demonstration contained in that work.

It being admitted that relationship was introduced by mental contemplation of the facts of blood-connection, is there room for the suggestion that relationship through the mother escaped notice? Surely not. The blood-connection with the mother is more apparent than the blood-connection with the father, and, if there was a time when circumstances prevented the male parent from being "individualised in the mind," it would during that time have to itself all the mental contemplation of which men were capable. If we trust to inference we must conclude that the kinship flowing from common maternity would have the first chance of being acknowledged. And we know as a fact that numerous peoples have got this kinship without ever getting beyond it.

for, is its denial of relationship with a family to the descendants of its women. Now, all we are told as to this in *Ancient Law* is, that relationship was, in the primitive view, exactly limited by Patria Potestas; that—while, if a woman died unmarried, she could have no legitimate descendants—the children of a married woman fell under the Patria Potestas, not of her father, but of her husband; and that such children were thus lost to the family of their mother, and "outside the limits of archaic kinship."

But it need not be said that within times known to us, mere difference of Patria Potestas did not exclude persons from relationship with one another. If it had done that, the children of orphan brothers would not have been each other's relatives—whereas the children of all males descended from the same ancestor were each other's Agnates, notwithstanding that each family formed by them was subject to a separate Potestas. How then, in the case of descendants through males, came mere difference of Patria Potestas not to have the effect of restricting relationship, if it is enough (it is all we are told of) to account for the exclusion from relationship of all descendants of married women? The primitive view, by assumption, treated men and women alike—the reason which excluded women and their descendants from relationship sufficing to exclude men and their descendants also. Did Patria Potestas, in the advance from the primitive condition, learn to discriminate between descendants through males and descendants through females, and to treat them differently—so that it made no bar to relationship between the former, while excluding the latter from relationship? It would seem that this must have happened. And, if so, how did it happen? At any rate, we have attributed to the Potestas exercised by the husband of a female an effect which Patria Potestas did not have (or did not continue to have) when exercised by male descendants of her family. And it was indispensable that it should have been explained to us how Patria Potestas could have an effect in the one case which it had not in the other. *Ancient Law* gives no explanation of this; and that means that it does not enable us to see how it may have been that Patria Potestas was the foundation of Agnation.

It appears, therefore, that, instead of a theory of Agnation with some claim to be considered plausible or satisfactory as a hypothesis, *Ancient Law* comes very far short of giving us a hypothesis at all; since it leaves both the inclusions of Agnation and its exclusions unaccounted for. What it gives us as an account of Agnation is simply the suggestion that Agnation was "as it were, a mould" which retained the imprint of the paternal powers after they had become extinct; and the suggestion that, in primitive times, when Patria Potestas limited relationship, the children of a married woman would be under the Potestas of her husband, and not under the Potestas of her father. It has already been shown that no help in inferring the origin of Agnation can be had by means of the former suggestion. It remains for us to see what is to be said as to the validity, on Sir Henry Maine's own view, of the latter suggestion.

Admitting, then, that Patria Potestas moulded relationship, and that, on the primitive view, relationship was exactly limited by Patria Potestas, it cannot be valid in any degree or to any effect, except on the assumption that marriage always, or as a rule, involved a change of Potestas; that marriage carried the bride away from her Paterfamilias; that marriages did not occur between persons who were under the same Potestas—did not occur, that is, between persons who were of the same *relationship.* This, however, is a large assumption. It requires to be justified; and, if it is not justified, *Ancient Law* fails to account—even to the small degree to which it has tried to do so—for the exclusions of Agnation. It need not be said that this assumption has not been justified in *Ancient Law.*

As to the need for justification, surely it should have been shown to us what there was to prevent the marriage of persons living under the same Potestas—or, to use the more suggestive equivalent, persons of the same *relationship.* We have the primitive Paterfamilias represented to us as governing his "children, grandchildren, and great-grandchildren," besides persons adopted by him and their descendants; and, therefore, as having under his Potestas a number of marriageable men and women. Marriages might therefore occur within the Potestas or relationship, unless there was something to prevent them. They might even be frequent. And, if they were anything like common, then, in the primitive relationship—to which we are referred for the basis of the relationship ultimately acknowledged—a man's descendants through males would be his descendants through females also to such an extent that that relationship could not, it would seem, develop into Agnation. It was necessary then to show us that persons of the same Potestas or relationship did not marry one another—which can only be done by showing us some reason why they should not marry one another. And it was the more necessary that this should have been attempted because there is no lack of instances of marriage in early times between persons of the same family or who were each other's near relations. To take one which has been mentioned already, the story of Abraham shows us brother and sister-german marrying. And, in Abraham's family, we find also a marriage between uncle and niece—between a man and his brother's daughter. Compared with such cases, the marriage of more distant relatives, such as cousins-german, seems natural and easy. And yet what *Ancient Law* assumes, in assuming that marriage always carried the bride into a strange Potestas or relationship, is that no such marriages took place within the Potestas.

It need scarcely be added, as to the children of women who had not been given in marriage—of whom Sir Henry Maine remarks that they would be illegitimate—that apparently there could have been nothing to take them, at any rate, out of the Potestas of the mother's father.*

* Among the Hindoos the child of an unmarried woman belonged to her father, and was among his heirs.

As to the possibility of justifying the assumption that there was no marriage within the Potestas, we know of prohibitions of marriage only in connection with the acknowledgment of relationship founded on blood; and then as applying, not to persons living under the same chief or of the same group, but to all persons counted to be of the same kindred whether living in the same group or not. And if ever there was no relationship except that which could be made by subjection to the same family chief, the question why there should have been either prohibition of marriage between the persons connected by it, or abstinence from intermarriage among them, will be found extremely difficult to answer. The answer, which, of course, could only be conjectural, would have to be consistent with, or reconcilable with, the class of rather formidable adverse facts one or two of which have been mentioned.

It remains to point out that it is entirely by means of this assumption that Sir Henry Maine, in *Ancient Law*, makes any approach he has made towards accounting for the exclusions of Agnation. For Patria Potestas—supposing it somehow capable of moulding or "founding" relationship—would not have founded Agnation unless women, when they married, had passed into a strange relationship. It is not Patria Potestas, therefore, but a system of not marrying within the *acknowledged relationship*, which appears, on the face of his own exposition, to have been the determining cause of Agnation. It need scarcely be said that a custom which forbids marriage between persons of the same relationship is well-known under the name of exogamy.

Of the further question, how Agnation could survive Patria Potestas, and even survive it for ages, Sir Henry Maine has nowhere taken any notice. But it was indispensable for him to show us how this could happen. A married woman's children were lost to her family in respect of her husband's Potestas, and of that only—this is what he tells us. Why then should the descendants of married women continue to be lost to their families when husbands ceased to have Patria Potestas? The fact which had determined their exclusion from relationship gone, the exclusion should have ceased—unless, indeed, there was by that time at work for it some other influence which by itself was of force enough to keep it up. If there was any such influence it should have been indicated, and shown, by sufficient evidence, to have possessed the capacity attributed to it. The effect of this having been neglected is that the theory is left unfit to account for what is declared to be by far the most numerous and most important class of cases of Agnation.

Here we must repeat that Sir Henry Maine can show us Patria Potestas and Agnation together in one case only, and that, in that case, Patria Potestas survived Agnation. In that case, too, Agnation was broken down—Patria Potestas still surviving—by the circumstance which could most plausibly be thought of as likely to keep it up after Patria Potestas was gone. At Rome, Agnation was destroyed by its relation to the law of inheritance.*

To the preceding criticism—which, if valid, if only in the last branch of it, shows that Sir Henry Maine has entirely failed to make out, even by way of hypothesis, any connection between Patria Potestas and Agnation—a word or two should be added as to the view of primitive relationship from which, in *Ancient Law*, he sets out. That the relationship of primitive times (if it is to be so called) was confined to the family, to those who were actually living under the same family chief, and that the primitive father was an unsocial being who acknowledged no ties except the tie of power which bound his dependents to him—is, of course, sheer assumption. And all we know of man's nature and ways makes it an exceedingly improbable assumption. It is from an assumption derived from a supposed pre-social state of man, however, that *Ancient Law* leaves us to grope our way towards the relationship acknowledged in the rather artificial Roman society.†

* It may occur to some people that, the kinship once established, the force of custom might keep it up. But custom tends to support everything that is established. It would tend to support Patria Potestas as well as to support Agnation. How can we know that it would support Agnation after it had let Patria Potestas go down? There is only one way of knowing—we must be able to show, by means of cases in which Patria Potestas and Agnation occurred together, that it actually did so.

† In such a state of man as is supposed in *Ancient Law* there

Sir Henry Maine's essay, nevertheless, yields us a most valuable hint, if we go on to search for a better account of the origin of Agnation.

For, as has been shown, he tacitly assumes that marriage would carry a woman out of the "relationship" of her birth, out of the "relationship" to which the male members of her family continued to belong, and that her children would be born in a strange "relationship." And if Patria Potestas ever excluded the descendants of females from relationship, it is quite clear that it was through marriage carrying women out of their original relationship that it got the chance of doing so.

Then it is beyond question that Patria Potestas—whatever it may have done at first—did not, as time went on, have the effect upon relationship in the case of the descendants of males which is attributed to it in the case of the descendants of females. Difference of Patria Potestas did not disconnect the former from one another, though it is said to have had the effect of disconnecting the latter from all of the former. But, both at first and at all subsequent times, its effects in the case of either must have been the same. It follows that difference of Patria Potestas cannot have had the disconnecting or excluding effect attri-

could be no such thing as right. A father's power would be what he happened to be able to make it. However great, could it be identified with a recognised and legal right like Patria Potestas? There could no more be paternal right than there was true relationship in Sir Henry Maine's primitive family.

buted to it in the case of the descendants of females, unless it be in consequence of some effect of a lasting kind produced by Patria Potestas in its earliest days, when, by supposition, the relationship founded by it did not last.

Now the important part, as regards Agnation, of the effect attributed in *Ancient Law* to Patria Potestas in its earliest period is that it restricted relationship to the Potestas. For any approach to accounting for Agnation, then, it was found necessary to assume that a person could belong only to one "relationship." We have seen that it was also found necessary to assume that there was no marriage within the Potestas or relationship; that there was no marriage between acknowledged relatives.

Without both these assumptions *Ancient Law* could not even have made a beginning—which is all it has done—of accounting for Agnation. But, between them, they account for the children of a married woman being, in the early days when relationship was a transitory thing, cut off from the relationship of their mother's birth.

How, originating when they are said to have been originated, these conditions could manage to last needed to be explained, and has not been explained. But it is easy to see that they can give no real help in accounting for Agnation unless, adding to the theory of *Ancient Law*, we assume that they did last, their sphere of operation expanding as relationship expanded — or rather, unless it is assumed that, with true relationship in existence, people could belong only to one body of relatives; while at the same time marriage carried women out of the circle of relationship in which they were born, and to which their male kinsmen belonged, into another circle of relationship.

It is easy to see, too, that the two conditions we have now arrived at are capable of accounting fully for the main features of Agnation, and that without any aid from Patria Potestas.

For, through the operation of the latter condition —and that without any aid from the husband's paternal power, and whatever its degree—the children of a married woman would belong to the relationship or kindred into which she had married; and in consequence of the former—without any effect of their father's paternal power, and whatever its degree—they could belong to that relationship or kindred only. Through the joint operation of those two conditions, therefore—without the help of the paternal power, and whatever the paternal power might be—children would be cut off from the relationship to which their mother belonged by birth. A practice of not marrying within the circle of acknowledged relatives is, of course, a practice of exogamy.

We discover by the help of *Ancient Law*, therefore, that those two conditions seem likely to give a full and satisfactory account of the exclusions of Agnation. Given a body of people acknowledging kinship through males, and acknowledging relationship to one another, among whom no relationship was acknowledged outside the body (clan or gens), and who did not marry within it—whose marriage custom, that is, was exogamy—then there appears to be the strongest likelihood that their system of relationship would in time exhibit every feature of Agnation.

It would follow that Agnation arose, where it did arise, only among exogamous peoples who acknowledged kinship through the father; and only among those of them with whom relationship was restricted to the clan or gens.

And as to exogamy—the custom which restrains from intermarriage men and women who are counted to be of the same kindred, of the same acknowledged blood-relationship—it may here be pointed out that it is, in fact, a far-reaching law of incest; and that it is founded upon relationship, and that therefore relationship in some form was older than it. It can have had nothing to do with the beginnings of relationship. On the contrary, ties of blood—of one sort or another—must have been thoroughly acknowledged before a custom so extremely inconvenient (to say nothing more) could have been founded upon them.

Agnation, therefore, cannot, if exogamy was a factor in producing it, date from the beginnings of relationship.

If we go on to inquire whether that which has been suggested to us as the other condition precedent to Agnation—limitation of relationship to a body or group—is likely to have arisen in early stages of relationship or in some relatively late stage, there is one well-known fact which can help us. The most obvious relationships of all, and those which, therefore, had the first chance of being noticed, are those which arise through the mother.* And relationship, where it is counted through women only, is not limited to a group. It connects together persons in different groups. It follows the related person wherever he is. This shows a tendency in early relationship—in primary relationship—not to admit of restrictions; a capacity of transcending the group where necessary; a tendency to reach to every person who was within its conditions and could be traced, and to include such persons wherever they might be.

Now, what could there be to counteract this tendency where relationships through the father were acknowledged as well as the more obvious relationships through the mother? Unless there appears to have been something which could do so, the probable opinion is, that strict limitation of relationship to the group does not belong to the earlier history of relationship.

Could the father's relation to his family have had a limiting influence upon relationship? Here we have to consider that, man being a social creature, there must have been bodies of people living together before there was any thought of relationship. This being so, the father's power over his family could have no chance of restricting relationship to a group unless the introduction of kinship through males at once dissolved the group into which it was introduced, and made every

* See *Studies in Ancient History*, by J. F. M'Lennan. London, 1876: "Primitive Marriage," chapter viii., page 121 *et seq.*

man in it the nucleus of a new group. That seems an extravagant supposition. But, even with it, it is most unlikely that the father's power, which could only for a brief period extend to the whole of a group, should permanently limit relationship to the group. Nor is it in the least intelligible why the father's power should cause denial of any actually perceived relationship.

A more promising supposition—that which appears indeed to be the most promising—is that the cohesion of the family, of the body of known relatives, of the whole group, of which kinship through males permits, would tend to make the group relationship the practically important part of relationship; while, on the other hand, relationship outside the group might even be not traceable beyond the nearer degrees. But, between that and the denial of all outside relationship there is a great difference. And, when we ask why this should follow, there appears to be again a want of reason.

On the other hand, this limitation can be understood if recognition of kinship through males did not take place until relationships previously acknowledged had had time to grow into a system, and the system had operated so that people were divided by it into kindreds, and that a person could belong to only one of them. The system of kinship only through females has thus operated. It has counted as one kindred all persons related to one another through women—all the persons whom it could show to be each other's blood relations. At first there was no relationship acknowledged outside this kindred, because there was none perceived. And, by the time that other relationships were perceivable, it had hardened into an institution, with conditions which united those who were born of it to one another and cut them off from other people; it was fully established as the social unit, and the primary social obligations depended upon it—obligations which no person could throw off without becoming an outcast; and which bound every person to one kindred exclusively, and to his own kindred as against all other kindreds; and thus—though, as time went on, with indications of approaching change—acknowledgment of relationship continued to be limited to it.

If the limitation of relationship to the group can be more probably explained in the way just indicated than in any other, we should have to conclude that Agnation, where it happened to be produced, was arrived at through a progress, set in motion by the recognition and assertion of paternity, from the system which counts kinship through females only.

Exogamy is constantly found with the system of kinship through females only, and, therefore, on this view, it would be antecedent to the limitation of relationship to the group. As it is found with kinship through females only, the father's power can have had no share in producing it. Reason has already been shown for the opinion that the father's power cannot have limited relationship to the group. Nevertheless, on the view of the history of that limitation here suggested, it can be made probable that the conditions which would favour Agnation would also favour the power of fathers over their wives and children.

It should be added that, while exogamy has been common among peoples who acknowledge kinship through males, the known cases of Agnation have been exceedingly few; and unless, notwithstanding this, there appear reason to conclude that it has been common—or, as Sir Henry Maine supposes, universal—if the limitation of relationship to the group was a condition precedent to Agnation, it would follow that this limitation has been rare. That both should have been exceptional—and there seems no reason for doubting that they were so—is intelligible on the view that kinship through males had to supplant in kindreds already established a pre-existing system of kinship. When this had been done, people might stop in general a good way short of complete denial of the relationships which had been supplanted.

It is perhaps unwise to attempt what—the limits and objects of the present work considered—cannot be done satisfactorily; but it seems worth while to develop more fully, even though in an incomplete way, the view of the origin of Agnation which has just been shadowed forth.

CHAPTER XIII.

THE ORIGIN OF AGNATION.

ROMAN jurists are agreed that, at first, the Patrician familia was identical with the Patrician gens, and *agnatio* the same thing with *gentilitas*; though afterwards when the gentile bond had grown feeble and the family bond had grown strong, and connections had come to be formed between gentiles and clients, the term familia came to be restricted to bodies of kindred within the gens, and it became necessary to discriminate between *gentilitas* and *agnatio*.* If this view be accepted—and there appears to be no reason against accepting it—since *gentilitas* and *agnatio* were originally the same, it is obvious that it is the gentile bond we ought to look to when seeking an explanation of the peculiarities of Agnation.

Of the gens thus much is certainly known—it was a body of persons having a common name and common religious rites, and considered to be related to one another. And very little besides is certainly known of

* "Family" is the common term nearly all the world over for a "tribe of descent," or stock-group indicated by a common totem.

it. It was generally believed to consist of persons descended from a common ancestor; but this was only a theory to account for the connection in the same body of persons who were considered relatives and had a common name and worship. Children belonged to the gens of their father; and it was natural therefore—the gentiles being considered to be unquestionably relatives—to suppose that there was a first father from whom they all were descended. But relationship throughout the gens could not be traced, and, though eponyms had been formed for most of the gentes, it is undoubted that the common ancestor was in every case mythical. The gens was really of unknown origin, of unknown history.

A common name—usually taken from an animal or plant—a common worship, or common object of reverential regard, and a bond of inherited relationship denoted by these are the marks of totem kinship (of which more by-and-by). And an alternative theory of the origin of the gens is that it was a body of totem kindred in which gentile connection had come to be taken from the father. It can be shown of the majority of the Roman gentes that they have every mark of having been groups of this type.

There are indications that the gentes were exogamous, but it must be owned that the evidence to that effect is not direct and not distinct. Plutarch tells us what, with the knowledge of early custom we now possess, can be easily received, that in early times the prohibition of marriage extended as far as the tie of blood;* and, if this be received, it involves—since the gentiles considered themselves to be of the same blood—that there could not be marriage between persons of the same gens. That, among the early Roman marriages recorded, there is an absence of marriages between persons of the same gens (this seems to be the fact), though much is not to be made of it, is consonant with this; for it tends to show that the gentes were exogamous in practice, whether required by rule to be so or not. Exogamy, however, in its widest sense of a prohibition of marriage between persons of the same stock-connection or kindred, has been so common that we may, upon imperfect evidence, venture to think that the gentes were exogamous. Even the direct assumption that they were exogamous would not be a violent assumption. That there was anciently extensive prohibition of marriage between relatives is clear enough, and—notwithstanding that we have been told that, in ancient times, people could not be relatives of their mother's relatives—at Rome prohibition extended to relatives on the mother's side. The first marriage of a Patrician at Rome within the seventh degree of cognation excited a sedition in the city.†

* Plutarch, "Roman Questions," vi.

† See "Anecdoton Livianum," in *Hermes*, vol. iv., pp. 371-2, Berlin, 1870. "Livius libro vicesimo. P. Celius patricius primus adversus veterem morem intra septimum cognationis gradum duxit uxorem. Ob hoc M. Rutilius plebeius sponsam sibi præripi novo exemplo nuptiarum dicens seditionem populi concitavit adeo ut patres territi in Capitolium perfugerent."

Paul Krueger and Th. Mommsen append to this short articles

Let us take it, then, hypothetically that the gens was anciently exogamous, and that the gens and the familia, gentilitas and agnatio, were anciently identical. And now let us see what has happened in the numerous cases in which, with exogamy for marriage law, kindred or family (gentile) connection has been taken, not from the father, as among the Romans, but from the mother.

In these cases, relationship being transmitted only through women, those persons only were considered to be relatives who were connected (or believed to be connected) with one another through female descents; and, in addition to the prohibition of marriage, which extended as far as relationship extended, there arose out of relationship another important consequence. The blood-feud united those who were considered to be relatives to one another, for mutual assistance and defence, and for revenge of injuries; and it helped powerfully to bind them to one another, and to sever them from other people. A man could belong—even after circumstances allowed of the fact of paternity being perceived and contemplated—only to that one kindred of which he was born, to which he was bound by the blood-feud, within the limits of which marriage was forbidden to him.

And, since a man could marry only a woman of a different kindred from his own, and children were of the kindred of their mother, a man's children were never of his kindred—they were "lost to his family." There

containing references to corroborative authorities regarding forbidden degrees of cognation.

could be no extension of a kindred through a male; it was the wife's kindred that was extended by marriage. We find in these cases—the family in the sense of stock or kindred being, however, the unit spoken of—everything that Sir Henry Maine has said of Agnation reversed as between the sexes—that it has all to be altered by the substitution in proper places of male for female and female for male. Instead of a woman being the "terminus of the family" (*mulier finis familiæ*), a man was the terminus of the family. A male name "closed the branch or twig of the genealogy in which it occurred." None of the descendants of a male were included in the notion of this relationship. It is clear that the system of counting kinship through females only—the custom which made children be of the kindred (gentile) connection of their mother—and the prohibition of marriage between persons of the same kindred produced those consequences without any aid from the powers which mothers possessed over their children.

And is it not suggested to us that—all other things remaining the same—a change in kinship which would make children of the kin or gentile connection of their father would yield for the body of kindred (*familia* or *gens*) the statements that would be denoted by Agnation, when equivalent to *gentilitas*, without aid from the powers which fathers possessed over their children?

To proceed, however, a man's nearest acknowledged relatives are everywhere his nearest heirs; and in general acknowledged relatives only are anywhere a man's heirs.

We find, accordingly, in the more advanced communities in which the system of counting kinship only through females, and exogamy have prevailed, that a man's own children, not being considered his relatives, did not inherit from him or from his family. Rights of inheritance arose to children in the families of their mothers. A man's nearest relatives were, first his brothers by the same mother, and then the children of his sisters by the same mother; and they were his heirs in that order.*

Here we have—for the family proper—a rule of inheritance which is the counterpart of Agnation.

And is it not suggested to us that change sufficient to make children of the gentile connection of their father, would in time yield the rule of inheritance which was the substantial thing in Agnation?

It is now, indeed, easy to see how, it being taken that in early times the familia or gens was exogamous, and *agnatio* and *gentilitas* indifferently expressed the gentile relationship, every feature of Agnation can be fully and satisfactorily accounted for, without any other condition than that relationship was limited to the

* Among the Ashantees, according to Bowdich, a man's son might inherit on the failure of sisters' children. Failing a son, the heir was the chief slave. The son, that is, on the failure of the proper heirs, was preferred to the chief slave. But in the Fantee country the chief slave excluded the son. The Fantees and the Ashantees were really the same people; and the place given to the son in succession among the latter must be taken as indicative of a transition in kinship. It should be said that we have no direct evidence of exogamy among the Ashantees. But there are many circumstances from which it can be inferred.

familia or gens as, with the system of kinship through females only, it is limited to the kindred.

For by effect of this restriction of the gentile relationship (*gentilitas* or *agnatio*), a person could belong to one gens or familia only, and could not have relationship with persons in any other. And as, by effect of exogamy, a woman could not marry a man of her own gens, and as children belonged to the gens of their father, the children of a married woman would not belong to the gens of her birth, and therefore could have no relationships in it. They would be lost to the familia or gens of her birth, and lost also to the sub-family in it, or family proper, to which she belonged by birth. A married woman would thus be *finis familiæ*—the terminus of the family, both in the larger and in the restricted sense of the word; neither the body of kindred nor the family proper could receive any extension through its women; the children of a married woman would have no right of inheritance in the gens of their mother's birth, or in the family in it to which she belonged—being by their own birth strangers to both—and all this without any reference to the powers, greater or smaller, which male parents might possess over their descendants. When bodies of relatives who could trace their relationship to one another (*familiæ* in the later sense of the word) grew up into distinctness and importance within the gens, their relationships also would be subject to the limitations of the gentile relationship—would be what the gentile relationship determined them to be; and the children of a married woman would be cut off from relationship with the familia to which their mother belonged by birth, and of course—since only persons who are considered to be a man's relatives can be his heirs—from all rights of inheritance in it.

And, since a man's nearest relatives are everywhere his nearest heirs, and children, when acknowledged as relatives, are evidently nearer than brothers, a man's heirs would be first his sons and possibly his daughters—but not married daughters if marriage passed women into another gens, or if women were settled with on marriage; next, his brothers by the same father; and after these the more distant members of his familia. It would not be surprising, but might rather be expected, that the body of the kindred—there being between all within it admitted though untraceable relationship—should come in as *ultimus hæres.*

Rules as to relationship and inheritance would thus arise which would present all the features of Agnation. They would be established while the gentile system was in full vigour, but there would be nothing to disturb them, as rules for inheritance, while it was gradually waning, and, as such, they would then be gaining in prescriptive force. Nor would it be possible to question them as rules for inheritance until what was most distinctive of the gentile system had disappeared, until, gentile connection having ceased to be of importance, the limitation of relationship to the gens had become obsolete. Even then, that they had immemorially formed the settled usage for the distribution of inheritances would give them a hold upon men which might not be easily shaken. They could not permanently survive the conditions which produced them; they were sure to give way under the dissolving influence of admitted relationship. But they might survive those conditions long.

This account of Agnation is consistent with the history of the breakdown of Agnation at Rome, which the account of it which derives it from Patria Potestas is not. The gentile system had ceased to be of importance before Agnation was seriously attacked; while Patria Potestas survived Agnation. And, considering that the conditions from which Agnation is here derived, with the difference of gentile connection being taken through the mother instead of through the father, unquestionably account for the exclusion of persons connected with a family or a kindred through males only from relationship with it, it seems difficult seriously to doubt that they give us—so far—the true explanation of Agnation. It is scarcely necessary to recall that there is authority for believing that *gentilitas* and *agnatio* were in early times coincident, and that there is some evidence to show that the gens was in early times exogamous. These points admitted, it seems clear that, to account for Agnation, we must further believe—and there is supplied us, in the parallel operation of kinship through females only, sufficient reason for believing—that it was preceded by relationship strictly limited to the gens, just as with kinship through females only, relationship is limited to the kindred.

When we go on to consider the conditions which appear to have produced Agnation, it may be well to point out that, if we try to show that Agnation was arrived at by a progress from relationship through female descents, exogamy, as a factor in accounting for it, is nothing but an aid. For exogamy is found with kinship through females only, and it is clear that, with that system, it took the leading part in marking kindreds off from one another. That the feeling against marriage between persons considered to be of the same kindred thus established maintained itself when new relationships had come to be acknowledged, accounts for exogamy being found with kinship through males. On the other hand, if we try to explain Agnation as a primary system of kinship, we have as a preliminary to account for exogamy—to show how it would arise where kinship was acknowledged through males.

As to the limitation of relationship to a group, some reasons have already been given for thinking that it did not belong to the earliest period of relationship. But that is beyond doubt if it has to be admitted that Agnation arose in exceptional cases only, and if it be as an exceptional form of kinship that Agnation has to be accounted for.

Now, as a fact, whether in ancient times or in modern, cases of Agnation are scarcely to be met with. It will appear hereafter that Sir Henry Maine, though he thinks it has been almost universal, has never been able to find it anywhere except in Rome. And there is no reason to think that it ever was common. On the contrary, there is reason to think that it always has been uncommon; for, to judge by the lasting power shown by the system of kinship through females only, to judge by the lasting power shown by Agnation itself at Rome, it would have lasted well wherever it prevailed, and, had it once been common, would now and then have been heard of. An approach to certain effects of it could not but be made wherever the clan or kindred had anything like coherence. That they stood together for mutual defence and for revenge was enough to bind the men of a kindred to one another, and to give the group relationship the first importance; and, even apart from that, it was inevitable that men should cleave to the body in which they were born, in which they lived, and with which most of their rights, duties, and interests were bound up. It is intelligible, too, that, where a kindred was settled upon land, it should deny to the outside relations of individuals any interest in its land; and that, in so far as the interests of its members were conjoined or common, it should deny to the outside relations of individuals any right to become sharers in them. But all this and similar things have happened, without denial of outside relationship, and with the recognition of effects attaching to it—without it being at all involved that relationship could be extended only through male descents—in other words, without Agnation.

And if it be as an exceptional form of kinship that Agnation has to be accounted for, whatever its early history, it cannot have been primary. If, while relationship was still young, kinship through males, when it began to grow into a system, had confined relationship to male descents through some influence emanating from it which acted immediately or speedily, Agnation, instead of being exceptional, would at first have been as common as kinship through males. And we may be sure that, in that case, it would long have continued common. For there would have been the influence which produced it to maintain and establish it; it would have had prescription in its favour; and the circumstances of early tribes would have tended to save it from serious attack.

Though not primary, however, it may conceivably have been arrived at otherwise than by a progress from the system of kinship through females only. It may have been produced among peoples who from the first had a system of kinship through males by influences which were not general, or which, at any rate, did not often prevail, which came into operation among them as time went on. But there are facts and considerations which favour the view that male kinship in general has been arrived at by a progress from relationships through female descents. The more important may be indicated.

That relationship happened to be reflected upon when the fact of paternity was obscure or uncertain seems to be the only possible explanation of kinship being in any case counted through women only; and it seems to be the only possible explanation of this kinship being anywhere developed into a system that, where it was so developed, the fact of paternity continued long to be obscure or uncertain. If this be so, what might be expected to happen when, where a system of kinship through women was established, the fact of paternity became clearly observable, so that it might be the subject of reflection? May we not feel sure that it would be reflected upon, and that, sooner or later, by one process or another, effect would be given to its suggestions of relationship? That the establishment of a system of kinship upon which kindreds had been moulded would retard the acknowledgment to practical effects of the new relationships need not be questioned; on the contrary, one might be prepared to find that, among peoples governed by custom, devices would have to be hit upon whereby to procure for them practical recognition. But when reflection, which had previously established a system of kinship through the weaker parent, had shown that there was kinship through the stronger, we need not doubt that means would in general be found of ensuring the recognition of this kinship. And, once recognised, it would, almost as a matter of course, become forthwith the more important of the two.

It is probable, in short, that, wherever a system of kinship only through females had been established, there would be, sooner or later, a progress from it to a system of kinship through males, and that the latter would thereafter become the predominating kinship.

It is obvious, too, that this progress may have taken place frequently, commonly, even universally, and evidence of it having taken place be wanting, or not readily recognisable for what it is—while the facts about kinship, superficially regarded, might seem to support a totally different view of its history. Transition facts disappear. In general, we cannot expect to have any record of them; and, even when we have, their significance is not always above dispute.

If transition facts can be found, however, and their significance be established, they will form a class of facts which really give us light as to what the history of kinship has been in the cases which exhibit the conditions which are present in them—in cases in which the factors which appear in them have been concerned in the moulding of society. And there will be nothing to make against the indication they give us as to what the early history of kinship has been in such cases. They will be evidence—and the best evidence there is upon the matter—to show that, in similar cases, there has been a progress from kinship through females to kinship through males. And probability will incline to the conclusion which they suggest.

Now, in a few—a very few—cases, what appear to be unmistakably transitional stages of kinship, or steps on the way to a change of kinship, are found among exogamous peoples. They occur among peoples who have advanced very slowly, but it is only among such peoples we could hope to find them. If, however, they show that transition has taken place, this is enough, as the matter stands, to raise a probability that transition has taken place wherever we find exogamy with male kinship.

There is a more numerous class of facts—better

known, too, than those just referred to—which can be understood on the view that there has been transition from a system of kinship through females, and appear inexplicable upon any other view. That, for example, early Hebrew custom permitted marriage between such relations as brother and sister german, and uncle and niece (brother's daughter), brother and sister uterine not being permitted to marry one another, is intelligible on the view that, relationship through the mother being recognised, relationship through the father was not to all effects established. And it seems as if nothing else could be made of it. It need scarcely be said that relationship through the father was most fully recognised among the Hebrews at a later period. Hindoo marriage custom (according to the commentator, Kulluka, at any rate) also presents a fact which seems explicable only on the view of a transition having taken place, while showing conclusively as to Agnation that it was unknown among the early Hindoos. Among the Hindoos a man could not marry in the primitive stock of his father nor (on the authority of Kulluka) in the primitive stock or his mother. The gotra or primitive stock, it should be said, has been in times that are known to us large and scattered, connection with it being indicated only by a common family name. That marriage within it was forbidden on account of it being the primitive stock—that is, on the score of relationship—is quite clear. And it is not to be doubted that the prohibition of marriage within it could not have been introduced in times when the gotra kinship was regarded as very remote—not to be doubted that it draws back to the primitive period of clanship, when the gotra was comparatively small and comparatively coherent. Now Agnation denies the relationship of a woman's children to her brothers. And here among the Hindoos—it being to the last degree unlikely that the fact should have been invented, and certain that, if it was fact, it must have been of most ancient origin—we find the relationship of a woman's children to every person of her stock admitted, so that they were prohibited from marrying any person of that stock. This rule is absolutely irreconcilable with Agnation—Agnation cannot have prevailed where it came to be established. On the other hand, it is easily understood on the view that there had been a progress from the system of kinship through females only. On that view there would have been prohibition of marriage within the mother's gotra before connection through male descents was recognised and prohibition of marriage extended to those so connected. And it is an obvious explanation of this rule that the prohibition of marriage was continued for the old kindred, that of the mother, when it was established for the new kindred, the kindred of the father.*

* The text of Manu (Manu iii. 5) prohibits marriage only in the gotra of the father. As stated above, it is in the gloss of the commentator, Kulluka, that the prohibition is extended to the gotra of the mother. Kulluka, however, is a very great authority. It is not to be supposed that he made so sweeping an addition to the text without ample reason, nor without being sure that the validity of the prohibition would be acknowledged as readily in the one case as in the other. The prohibition of marriage in the mother's gotra, it may be

But, further, it appears that the family system amongst many peoples which have been exogamous is far more intelligible upon the view that it is the issue of a progress from a system of kinship through females only to a system of kinship through males than it is upon any other view. And there are single facts which seem almost to prove this—for example, the Levirate. The Levirate has prevailed far and wide. And, if of polyandrous origin, it carries us far back towards the social condition in which kinship could be counted through females only.

It seems worth while to add that we find in marriage among some of the greatest of ancient nations a custom which—while no doubt serving other uses also—would be effectual as a device for reconciling an ancient system which made children of the same kindred with their mother with the resolve that children should be counted of the family and stock of their father. The wife, at marriage, was taken into the family and stock (gens, gotra) of her husband, and admitted to his *sacra*, to participation in his religious rites; and she ceased thereafter to belong to the kindred and worship of her birth. Being, by this device, of her husband's family and stock, her children—even if they took gentile and family connection from her — would be of the gentile and family connection of their father.

said, is quite consistent with all we know about the Hindoos. It has been remarked already that, where Manu mentions father and mother together, he often mentions the mother first. And, in some of the most important passages in Manu relating to parent and children the text speaks of the mother only.

If the transition did take place, the problem for those by whose means it was carried out would have been: Given kindreds already formed, bound together by the blood-feud and the prohibition of marriage within the kindred—kindreds connection with which had been transmitted through female descents only—how to secure that connection with the kindred should come to depend upon male descents; in other words, how to secure that children should be counted of the same kindred with their father instead of being counted of the kindred of their mother. Men would have no reason for shaking themselves free of the existing kindreds; nor is it intelligible how they could do it if they wished. For the kindred would be the basis of society as it existed; acknowledged relationship would be limited to it, and the rights and duties connected with relationship also; every man would belong to one kindred, to which he was bound by the strongest ties at the time acknowledged. To shake it off could hardly even be thought of; and all that men would have a motive for doing, would be to change the kindred or gentile connection of their children—who, as matters stood, would not be the relatives of their own fathers—to wrest children from the kindred of their mothers, and to win them for the kindred of their fathers. That anyhow accomplished, there would be no obstacle to the full affiliation of children to their father.

It might be expected that various methods or devices for effecting that would be employed—and there are indications of this having been the case. Being of the nature of devices, they could succeed only because men would generally be favourable to the object for which they were used. And, though an effect upon the kindred of the children would be necessary to their succeeding, or would necessarily take place if they succeeded, it would be the individual members of kindreds by whom they would be employed. It is easy to see that it would be indispensable to them all alike that a woman's relatives should cease to have rights over her children. Perhaps it might be expected that they would frequently involve that the woman's relatives should give up to the husband the woman and her children, and cease to have to do with them. This could not be accomplished without contract. And it appears to have been in many cases accomplished by means of contract alone.

The most effectual method, no doubt, would be for the man to be able, in virtue of the contract he had made, to bring the woman into his own clan; and where religion was influential, probably this was always, in the long run, found necessary. This being done, a man's children—though taking gentile connection from their mother—would be, by birth, of his own kindred, among his acknowledged relatives, and there would be nothing to prevent him from having them for his children in the fullest sense. A device by means of which they would be, by birth, of his own gentile connection would make them his children. They would become his children through being by birth of his clan.

It is in a class of cases in which religious influence

was slight, and made little or no difficulty, that the connection with the mother's kindred seems to have often been got rid of by means of contract alone. The mother's relatives having been satisfied, their non-interference having been secured, the husband got the children for his own, and was free to count them of his own family and clan.

Whatever method was used, however, the gentile system being what it was, when a man won his children for his clan he would win them for himself, and when he won them for himself, he would win them for his clan. And, as the change went on, the gentile bond would apply the old restraints and obligations to the kindred as now recruited. What wonder, nevertheless, if, for a time at least after the change was generally accomplished, the restraint on marriage should extend to the mother's kindred as well as the father's? We see, indeed (by such marriages as those of Abraham and Nahor), that, in the case of the Hebrews, the restraint as to the mother's kindred was alone effective for a time. And, though children would be now of the clan of their father, and not of the clan of their mother, what wonder if, in general, relationship to the mother's relatives were to continue to be to some effects admitted? The gentile bond would still tend (so to speak) to deny relationship to those who were outside it—and, therefore, to lead up to Agnation. But the associations of the old relationship would resist it. The restraint on marriage, where that continued for the mother's kindred, would, of course, be an admission of relationship outside it.

The more completely children were severed, by the circumstances surrounding marriage, from the mother's kindred, however—and, marriage everywhere tending to fall into some convenient uniform fashion, the kindred would be recruited through all the men contained in it under much the same conditions—the more tightly could the gentile bond be drawn. Where the resistance to it was slight it would succeed, and the kindred would be on the way to Agnation. What the circumstances were which would favour it may perhaps be gathered if we bear in mind what Roman marriage was.

It is obvious that the elaboration within the present work of the view now presented is not to be thought of. But some further consideration of the conditions which may be believed to have affected early kindreds seems almost necessary, and may perhaps be ventured upon. What, in a general way, these conditions were may be gathered by observing the conditions affecting those rude societies which are known to us. It will be convenient to take for examples societies which have been observed in modern times.

Wherever, with exogamy, kinship is counted through females only, there are at least two kindreds represented in every household—the husband is of one family or kindred and his wife and children are of another. And, if he has more than one wife, there may be represented in the household an additional kindred for every additional wife. An interfusion of kindreds in every household—one represented by the husband only—is in all cases inevitable.

In the simpler or ruder examples of this kind, however—those, for instance, which occur among the Australians or North American Indians—the degree of interfusion is very remarkable. It commonly happens that all the separate communities—local tribes, or, to use the North American term, nations—which occupy an extensive district, are, in the main, similarly composed, each being made up of persons belonging to the same—four or six or more—separate kindreds. The local tribe or "nation"—the body of people who live, and hunt, and make war together—is not itself a kindred group; it is anything but that. It is an agglomeration of persons belonging to several separate families or kindreds. Persons belonging to different kindreds are contained in every household. Persons belonging to four or six or more kindreds make up all the households in the community. Even a single household, when there are several wives, or when it contains three or four generations—every man marrying into another kindred than his own, and his children being of the kindred of their mother—may contain representatives of every kindred included in the "nation."

And, on the other hand, as has been said already, the kindreds which may be all represented in a single household in one community are commonly found to be the components of all the communities in an extensive district. These communities, moreover—the local tribes or nations—are not fortuitous combinations, or of recent origin. They believe themselves immemorially to have been composed much as they are at present.

The families or kindreds, which make up one and all of them, consist each of persons related to each other through female descents. Of course, their relationship is mostly untraceable. All of a kindred, nevertheless, acknowledge relationship, and the restraints and obligations of common blood. Whether in the same community or not, they do not intermarry with each other. And the blood-feud unites them for mutual assistance, reparation of injuries, and revenge. The latter obligation, however, is frequently confined to those of the kindred who are of the same community or local tribe.

How then are the members of a kindred so diffused recognisable by each other as belonging to the kindred? They have what serves them for a common name, but is much more than that. To use a convenient (though in some respects unfortunate) American term, they are of the same totem, and their relationship is indicated to them by the totem. The totem is some natural object—commonly an animal or plant, sometimes a heavenly body; and they figure it upon their bodies (whence tatooing), and call themselves by its name. All of one kindred are Bears; all of another Wolves; all of a third Suns, and so on; and the totem being taken from the mother, a stranger is at once recognised by those who bear the same totem mark and totem name with him as their relation connected with them (though in an unknown way) by female descents. The totem or kindred name is the more prominent because personal names are usually carefully concealed. It more than serves the uses of a family name. It is the totem mark alone that is placed upon a grave. It is with it alone that, in intercourse with Europeans, the representative of a kindred signs a treaty.

But, further, the kindred believes itself to be somehow *descended* from that (the animal let us say) which is its totem, and to be still mysteriously connected with it. They reverence it as their protector and friend. They have what must be called a religious regard for it. They consider an individual of the species as their relative, and call it brother. And they will on no account hurt, or kill, or eat it. Totem kindreds, at a somewhat more advanced stage—for it may be said that the totem can be traced almost everywhere—often try, by filing, distorting, or pulling out teeth, by arrangement of the hair, and other devices, to indicate by outward resemblances their connection with the animal which is their totem. Sometimes they believe that it gives them counsel in dreams. They are constantly found believing that they will take its animal form after death; and varieties of what may be called the were-wolf superstition also testify to their conviction that their human form is interchangeable with the animal form of the totem. When people become speculative, a great totem is often made out to have been the creator of all things. And then, strong as the objection to killing or eating it is, we find that frequently its body and blood are partaken of eucharistically. In the very rudest cases, however, the kindred—besides getting from the totem a common name and a test of inherited relationship—have in it the nucleus of a common worship.

Totemism is not our subject; but it cannot be amiss to recall that a hypothesis has been framed to account for the animal and plant worship of ancient nations on the view that they all passed through the totem stage;* it being shown, among other things, that that worship, in numerous cases, exhibited the leading notes of totemism; that animals, for example, were worshipped by tribes of men who were named after them and believed to be of their breed, and never eaten unless eucharistically. And on the other hand, it may be pointed out that the totem unquestionably carries us back at least to the earliest days of relationship—to a period while there was yet no prejudice against the marriage of kindred persons.

Scattered throughout various separate communities as the kindreds we have been noticing are, it is almost impossible to resist the suggestion that there was once a separate coherent nucleus for each. The communities are just what the interfusion—by exogamy (or capture) and the system of kinship—of so many originally separate, or at least coherent, groups, each of which considered itself a kindred, would make them. And it seems absolutely impossible that the totem bond should unite a kindred scattered throughout different communities, as it now does, unless it had marked it off from other kindreds before systematic interfusion of kindreds

* "The Worship of Animals and Plants." By J. F. M'Lennan. *Fortnightly Review*, 1869-70. See also "Animal Worship and Animal Tribes among the Arabs, and in the Old Testament." By Professor W. Robertson Smith, *Journal of Philology*, vol. ix.

had begun. Without it all relationship, except what memory could take note of, would now be lost sight of. While it yet was not, apart from memory, all relationship must have been lost sight of. It is practically all there is to preserve the tradition of relationship. We may take it, therefore, that such relationship as preceded it, if any, was lost sight of. And to follow the blood as it does now, it must have marked it at a common source. The totem therefore was precedent to that by which the interfusion of kindreds is regularly caused, exogamy; and there can be no trace left of any interfusion—made by the capture of women or otherwise—which preceded it. We are thus carried back to a time when there were bodies of kindred having each a common totem, which were necessarily bound together for common defence—that is, essentially by the blood-feud—which were connected by that system of relationship which first was formed, and which had no objection to the intermarriage of related persons.

Among the North Americans, and even among the Australians—not unfrequently among the former, only rarely among the latter—we find also totem kindreds in which kinship through males is acknowledged, and the totem or gentile connection is taken from the father. Where information about the marriage law is accessible, it is that the kindreds are exogamous—that is, that persons of the same totem may not marry; but we come also upon arrangements to prevent marriage between near relations who are not of the same totem. Of such cases, when there is nothing to suggest that there has been a recent change of kinship, the probable explanation is that a general practice of polyandry in the Tibetan form—with which there is certainty as to the father's blood, though not as to the father—at an early stage made a transition to kinship through males natural and easy. That there should anywhere be kinship only through the mother can only be explained by there having long been uncertainty about fatherhood; and in "Primitive Marriage," reason was shown for believing that that kinship must in numerous cases have given way before the certainty of the father's blood obtained with Tibetan polyandry.*

In some of the cases now being spoken of, however, there are indications that there has been a change of kinship at a period not remote. In America, for example, the Algonquins, on the whole, have kinship through males; but there are detached portions of that people—not very long separated from the main body—among whom the totem is said to be still taken from the mother. In America, at least, the circumstances of the family have long been not unfavourable to the acknowledgment of relationship between father and children. To prevent it there has been the force of custom—the fact that children have always been of the totem of their mother. But, once the desire to have their children for themselves and for their own totem became general among men, some way of giving effect to it was sure to be found; and we may be content with knowing

* *Studies in Ancient History.* By J. F. M'Lennan. London, Quaritch, 1876. "Primitive Marriage," chapter viii.

this where we find circumstances suggestive of a recent change.

One thing we can be confident about. It has been shown that the totem is older than exogamy. That, like the totem kindreds previously noticed, the kindreds now spoken of have acquired the prejudice against marriage within the kindred is enough to show that, up to a certain point at least—until after exogamy had been established—the circumstances and history of both were substantially the same.

The change in kinship, supposing it to have taken place in the cases just noticed, was sure to add greatly to the cohesion of the kindred, and proportionally to weaken the cohesion of the local tribe or "nation."

With kinship through females, in a household which contained three or four generations every totem in the community might be represented. The kindreds were so interfused that close observation was needed to show that they were not all one body of relatives. Family ties mitigated the virulence of the chief disruptive force, the blood-feud. And, indeed, the community could not be broken up without a break-up of every household included in it. With the new kinship all of a household, except the wives, became of one totem, and were one for the purposes of the blood-feud. It was now natural for the households that were bound together by the blood-feud to draw together and form separate bodies within the community. That done, they were totem groups again—proper clans or tribes of descent, too —with nothing except the conditions found to be indispensable for living with one another to check the disruptive influence of the blood-feud. And this is what we usually find with clanship where, gentile connection being taken from the father, the community or local tribe is made up of a number of separate clans.

Since we have seen that, with the earlier kinship, the same kindreds often enter into adjacent communities, the view that there has been a change of kinship is greatly favoured by our finding, as may often be done, that the same clan is a component of more communities than one—is supported by it in proportion to the difficulty of otherwise accounting for such a fact. And it is favoured also by our finding—as may be done in Australia—the segregation of kindreds from each other within the same community, though far advanced, not yet complete. In the latter case it may be inferred that a change of kinship took place at no very remote date.

The kindreds in which a change in kinship took place within the totem stage, when they had thought their totem — the mysterious parent, protector, and friend—into a god, would, with their common name and common worship and tradition of relationship, greatly resemble the totem kindreds in which countries like ancient Italy abounded. In very many cases, however, the totem seems to have developed into a god before the change in kinship took place. Then, each kindred holding stiffly to its own worship, it can be seen that it interposed a formidable obstacle in the way of the change (which, on the other hand, it supplied a new motive for making)—an obstacle which it taxed the ingenuity of mankind to surmount. We can see very well among peoples known to us how great the obstacle was, and we can discern among them also what were some of the expedients by which it was surmounted.

In Western Africa we find, with circumstances seemingly favourable to the acknowledgment of relationship between father and children—the men masters, severe laws against adultery, and a certain accumulation of wealth—the only drawback being an unusual frequency of polygamy—the system of kinship through females only still firmly rooted; and, though polygamy may favour it, it is religion chiefly which keeps it up. The kindred to which a person belongs determines what god he shall serve, what religious observances he is bound to, what is to him forbidden meat, and—almost certainly—what women he may not marry. And, in marriage between equals, or when the conditions are of the usual sort, the force of religion sustains ancient custom so that children belong by birth to the kindred, and to the god and worship, of their mother. Wives, it may be said, are purchased; but, if they have property, when they marry the husband does not get it—it remains the wife's and goes to her children, who are heirs, not of their father and his relatives, but in the family of their mother.

In Africa, contract—that is purchase—appears to have been found, in a class of cases which will be looked at hereafter,* of itself sufficient to carry children

* See Chapter xvii., on "Sonship among the Hindoos."

to their father from the kindred of their mother. In the cases now spoken of, though used, it was not found sufficient. But it is very curious and instructive to find among the people of Guinea a custom which shows that there, unaided from without, men were in a good way for getting over, by the help of religion, what religion had made a formidable difficulty. It has been shown that the custom found among ancient nations of admitting the wife at marriage into the kindred of her husband, and to participation in the sacred rites to which he was bound, would be effectual as a device for reconciling the ancient system of kinship through females with the resolve that children should be of the kindred of their father. Now we find this very device in common use in Guinea; and, though the purpose for which it was used is not clearly stated, and it may have been intended to serve more purposes than one, there can be no doubt as to one effect it was sure to produce. It could not fail to make children—in the cases in which it was employed—of the kindred and worship of their father; for, as it was used, there was no other kindred or worship they could belong to.

In general, Bosman tells us,* marriage carried the consequences already described. In the case of his chief wife, an equal, the husband perhaps could not hope to escape from them; and, even in his ordinary marriages he appears to have had to put up with them. It was customary, however, for a man to buy and take

* Bosman's *Description of Guinea.* Translated from the Dutch. Second edition. 1721. Page 169.

to wife a slave, a friendless person with whom he could deal at pleasure, who had no kindred that could interfere for her, and to consecrate her to his Bossum or god. The Bossum wife, slave as she had been, ranked next to the chief wife, and was, like her, exceptionally treated. She alone was very jealously guarded. She alone was sacrificed at the husband's death. She was, in fact, wife in a peculiar sense. And, having by consecration been made of the kindred and worship of her husband, her children would be born of his kindred and worship.

Whether intended or not, therefore, we find here a beginning made of kinship through males, or, at any rate, a great advance towards it—at least, an expedient found which could make children of the kindred of their father, the use of which, if its effect was welcome, was likely to spread quickly. It is now easier to believe that, in the other cases in which we know the same custom to have prevailed, one reason at least for it coming into general use was the effect which, by means of it, could be produced upon kinship.

It need not be supposed that, in those other cases, the expedient was first used in the case of slaves—though it is not unlikely that that was sometimes, or even often, the case. However that may have been, in a progress from the Bossum system as it was in Guinea, once a feeling in favour of the new kinship had become general and strong among men, Bossum marriage would cease to be confined to slave women; gradually men would get into the way of giving their own women in marriage on the new conditions. Once these had become the common conditions, the contract for marriage would commonly surrender to the husband (as we know that it has constantly done) all rights of the woman's family over her and her issue; while consecration to the husband's god, making her and her children of the husband's clan and worship, would free her family from gentile connection with and gentile duties towards them.

With Bossum marriage, as it was at first, moreover, children would have been, by the circumstances of the case, cut off from the relations of their mother. And with this form of marriage spread as has been supposed, there would even be present the circumstances which favour the rise of Agnation—the circumstances which would dissever a married woman's children from her kindred so as to oppose the least resistance to the old view of the gentile bond, which confined relationship to those who were within that bond. Nevertheless, the rise of Agnation would, in such cases, be far from certain. Much, it would seem, would depend upon whether the transition from the old kinship to the new was quick or slow—upon the length of time for which the old system maintained itself beside the new. For, with the transition protracted—and men's minds long kept fixed on the fact that there was relationship through the mother—there would be greater likelihood that some acknowledgment of that relationship might be thought consistent with the prevalence of the other.

Polygamy, therefore, which would obviously lengthen

and graduate the process of change—since it would be more difficult for a man to get all of many wives on the new conditions than to get one—which, moreover, by dividing a man's children into groups each specially connected with a mother, the persons in each being specially connected with each other because of their mother, in all cases emphasises the connection between child and mother and the connection which arises through the mother—and which would in both ways keep up the tradition of kinship through the mother alongside of the new kinship through the father, might, where it prevailed to any considerable extent, counteract the circumstances which make for Agnation, with the result that, when the latter kinship generally prevailed, there would still remain acknowledgment of kinship through the mother. With monogamy the change would be more rapid, and much more likely to be complete. It is scarcely necessary to say that Roman gentile marriage was monogamous, and that the bride exchanged the gens and sacra of her birth for the gens and sacra of her husband.

It may be well to add that where—as among the Hindoos—after male kinship had been established, prohibition of marriage remained for men as regards all women of the name of their mother, Agnation in anything like strictness could not be looked for. And, among the Hindoos, we find that a man, when he had no son, could—just as if kinship had been through the mother only—reserve to himself a son of his married daughter—necessarily by a man of another stock (gotra), and take him to be of his own name and stock, to be his heir and the continuer of his family.

Of the power which a husband possessed over his wife and children when they had been made fully over to him by the woman's kindred, and the protection of that kindred—as a duty, at any rate—had been withdrawn from them, it may be said that custom would everywhere settle what it was; and that it is only by a knowledge of custom we can know what it has been. Naturally it would somewhat vary. It would tend to be great. But, were it unlimited—and though we can find the paternal power recognised by law as unlimited in one case only, it would be very rash to say it never was unlimited in any other case—power possessed as a consequence of purchase could not be primitive. We should be put upon inquiry as to the history which had led up to it.

It should be noticed, too, with reference to the Patria Potestas theory of Agnation, that it was not the power (whatever it might be) which men acquired over their wives and children, but the fact that wife and children were cut off from the woman's kindred, that would favour the rise of Agnation.

In another class of cases—which may have been common—we find the transition from the one kinship to the other going on under circumstances less favourable to the rise of Agnation. These are cases in which, for a time, the gentile connection of children was not taken regularly from either father or mother, but sometimes from the one and sometimes from the other—it being settled, in some cases by the relatives before marriage, in others by the parents when the birth of the child was at hand, which clan a child should belong to. In all such cases we find the totem grown into a god, but with its totem origin clearly recognisable; we find the kindreds interfused much as they are when gentile connection is taken from the mother; and they all involve that the persons who married each other belonged to different clans.

We come upon a few cases of this sort in Africa, with evidence that, just as in marriages between Protestants and Catholics among ourselves, the boys belong to the clan of the father, and the girls to the clan of the mother. But the Hervey Islands furnish the best case of the kind.* In the Island of Mangaia, parents, at the birth of a child, arranged between themselves whether it should be dedicated to the father's god or to the mother's. The dedication took place forthwith, and it finally determined the child's gentile connection. And gentile connection carried with it obligations and liabilities such as arise with it when it is determined by the blood of the mother or the blood of the father. There is evidence enough to make very probable, what in such a case could hardly be doubted, that there had previously been kinship only through the mother. On the other hand, we are told that, on the whole, the father's clan and god got the preference; and as the father carried off the child to be dedicated immediately

* See "Myths and Songs from the South Pacific." By the Rev. W. Wyatt Gill. London: Kegan Paul & Co. 1878.

after its birth, and dedication was final, he always—if he chose to break the compact made, which sometimes happened—had the matter in his own hands. The Mangaians therefore were passing through a transition towards kinship and gentile connection being taken from the father. A little later, and all children would have been of the father's clan and worship. Later still, the wife, when she married, would probably have exchanged the clan and worship of her birth for the clan and worship of her husband.

In these cases, children of the same parents belonging to different clans, no clan containing all who were of the same blood on either father's side or mother's side, and every clan being made up of persons who, though attachable to it in right either of their fathers or their mothers, were bound together only by their common worship, and the obligations and liabilities which followed upon it, the clan—which had ceased to consist of persons believed to be connected through female descents, and was on the way to being composed of persons who would believe themselves to be connected through male descents—was, for the time, more like a religious body than a body of kindred. The sense of relationship on the side of both father and mother—though carrying no gentile effects—could subsist and develop with it. And, when finally the father's kinship and worship prevailed, the sense of the other kinship might well be strong enough to prevent clanship from becoming agnatic.

Enough has now been done to indicate how the circumstances of early peoples appear to have affected the tendency of clanship to become agnatic. Another class of cases in which there seems to have been a transition from kinship through females to kinship through males —effected apparently by means of contract alone—will be more conveniently considered in a future chapter.*

It scarcely need be added that all that has here been said has been said of cases in which exogamy, that is prohibition of marriage between all persons considered to be of the same kindred, became established as marriage law. That exogamy has prevailed most extensively is well known; it may be believed to have been general, and nearly universal. An extensive practice of polyandry might however (as has been reasoned in "Primitive Marriage") prevent the rise of exogamy.† Of such cases, in which the exogamous restriction never was established, all that need be said is that the religious and quasi-religious difficulties which have had to be noticed could not have occurred, or were very unlikely to occur, in them. Husband and wife might be of the same kindred; and, as soon as certainty of fatherhood permitted, no obstacles would lie in the way of gentile connection being taken from the father, and full allowance made for all ties of blood, whether deduced through males or through females.

* Chapter xvii., on "Sonship among the Hindoos."

† *Studies in Ancient History.* "Primitive Marriage," chapter viii.

CHAPTER XIV.

EXAMPLES OF AGNATION.

It seems unnecessary to repeat all that has been said in chapter v. to show that Agnation did not prevail among the early Hebrews, and that, among them, relations on the mother's side were kindred in the fullest sense. It seems enough to mention Laban's family, of which his married daughters, and their children, and their husband too, formed a part; and the case of Abimelech, who is acknowledged by his mother's kindred as "their bone and their flesh," and aided by them on the ground of relationship.* It may be well, however, to recall that we have met with what seems clear Scriptural indication of beenah marriage as prevailing at a very early time—a form of marriage with which Agnation is impossible, and which, on the contrary, cuts off the husband himself, as well as his children, from the connections of the husband's birth.

* Prof. W. R. Smith points out that in Lev. xxv. 49 (Heb.), "of his flesh" is explained to mean "of his clan." Thus it would seem that his mother's kin recognise Abimelech as their clansman. So, too, we find among the Australians that people of the same kindred may not intermarry because they are "of the same flesh."

And it should be added that marriages like those of Abraham and Nahor could not occur except at a time when relationship through males was not itself thoroughly enough acknowledged for a bar to marriage to have been raised upon it.*

To show the absence of Agnation among the earliest Hebrews known to us is all we are concerned to do—though, no doubt, if Agnation could be shown to have prevailed among later Hebrews, we should have proof that, as regards the Hebrews, Agnation was not a primary kinship; and also proof—since the Hebrews never had Patria Potestas—that Agnation could arise independently of Patria Potestas. But Agnation was unknown among the later Hebrews.

We learn from Numbers xxvii. 1–11 that, after the Israelites had become a people, when they had made conquest of territory and were just about to divide it, the daughters of a man who was dead put in a claim to share, as representing their deceased father, in the land to be allotted to the men of their tribe; and that the claim was at once admitted, and, in the first instance,

* With those cases in view, it seems as if there can be no reasonable doubt that the Athenian custom which, in the case of an heiress allowed brother and sister german to marry, became established at a time when, at Athens, the law of incest made no real difficulty for such a marriage. And with this it agrees that, according to a tradition preserved by Athenæus from Clearchus of Soli, a disciple of Aristotle (Müller, Fr. Hist. Gr. ii. 319, Athen. xiii. 2), the Athenians practised Nair polyandry before the time of Cecrops, and that no man knew who was his father. The story of King David's children, Amnon and Tamar, shows that marriage between brother and sister german was allowed among the Hebrews at the time to which it refers—without there being any property reason for it.

without any conditions—it being at the same time prescribed as a rule that the daughter should succeed her father in land, failing a son. A later passage (Numbers xxxiii. 1–9) tells us that a condition was subsequently imposed upon the admission of this claim, and upon the law as thus established; and this passage is extremely interesting and instructive as disclosing the motives for those rules upon which, though they disprove Agnation, a belief in the general prevalence of Agnation has mainly been founded. The chiefs of the tribe came forward and pointed out that the effect of allowing women to become sharers in the land of the tribe might be—since they might marry men of other tribes—that men of other tribes would acquire a settlement upon the land of the tribe and after a time carry it off to the tribes to which they themselves belonged. It was thereupon laid down that the claimants, as a condition of getting the share they had asked for, must marry within "the family of the tribe of their father." The law prescribed for the like cases was correspondingly modified, and, as so modified, it continued to be the law for Israel. A desire to prevent confusion of tribal lands—to keep the land of each tribe intact for the tribe—was the motive for the condition imposed upon the daughter's right of inheriting. Had the law of incest been so strict as to make the marriage of the daughter within the paternal kindred a thing not to be thought of, the same motive might have been of itself enough to account for her being excluded from inheritance altogether. In this case, be it observed, the right of a daughter to inherit is not said to have been questioned—though the inconvenience that might attend it in a new case is said to have led to its being subjected to a condition. When the son of an heiress came to inherit he obtained the inheritance in right of his mother.

It may here be said that we see from 1 Chron. ii. 34–36, that, among the Israelites after they had become a people, the family could be continued through a daughter. It is there said that Sheshan, having no son, gave his daughter to wife to Jarha, his Egyptian servant, and that she bare him a son named Attai. And through this Attai the family genealogy is continued—just as if he had been Sheshan's son or son's son. The names of women, as can easily be seen, are of very frequent occurrence in the Hebrew genealogies. It may perhaps be worth mentioning also that there seems to have been no case for the Levirate when a deceased elder brother, though without a son, had left a daughter.

The case of the Hebrews, apart from its direct bearing upon the prevalence of Agnation, and the place of Agnation in the history of relationship, is of the utmost value, as showing beyond question that, from finding rules of inheritance which postpone daughters and their descendants to sons and their descendants, we cannot infer the prevalence at an earlier time of Agnation where such rules are found. For the Hebrews, when they had become a settled people, postponed daughters to sons in inheritance;

and their earliest records clearly show that the earlier Hebrews were not agnatic. The importance of this will appear immediately. The case of the Hebrews is also valuable, of course, for the light it throws upon the origin of rules of inheritance unfavourable to women. A clan which acknowledged kinship through males was sure to be continued chiefly through its males; and, when marriage within the clan was strictly prohibited, it could, apart from special devices, only be so continued. Inevitably, therefore, when a clan had settled upon land its rule for inheritance was such as to keep the land for the men of the clan—so that daughters could have no chance of inheriting land if there were sons, and, apart from special devices, possibly no chance even failing sons unless they could marry among their own kinsfolk.

To pass now from the Hebrews, although Sir Henry Maine has told us that agnatic relationship is discoverable almost everywhere, his particular statements as to the existence of Agnation have always been very few. And, in his latest work,* his earlier statements have been most seriously modified. Let us briefly show what they were, and what they are, and what is implied in them as they now stand.

The Hindoos, with the Romans and Slavonians, are said in *Ancient Law* to furnish the greater part of the legal testimony to the truth of the Patriarchal Theory. And it is stated in that work (p. 150) that "in Hindoo law, kinship is entirely agnatic." In *Early Law and*

* *Early Law and Custom*, passim.

Custom it appears, however, that, from the earliest times known to us, a Hindoo who had no son might be succeeded, and was customarily succeeded, by the son of his daughter (the son of an appointed daughter); and we are told that customs akin to this appear to have been very widely diffused over the ancient world *—the daughter in all such cases, as Sir Henry Maine observes, becoming a channel through which the father's blood passed to a male child.

The modification made in the later work obviously involves an admission that, to judge by the evidence to which Comparative Jurisprudence usually confines itself, the early Hindoos—and also the numerous early peoples who were in the same case with them—were not agnatic. The two leading notes of Agnation are that a married woman is the "terminus of the family" —that the family cannot be continued through her; and that there can be no right of succession for any descendant of a woman. And, among the Hindoos, as Sir Henry Maine aptly says, a daughter could be, as it were, a channel through which the father's blood passed to a male child—that is, the family could be continued through her; and her child was her father's heir, and was in every respect as a son to him. What we know of the Hebrews makes it clear that from finding the right to inherit of a daughter or her descendants limited as it was among the Hindoos, there can be no inference of a prior prevalence of Agnation. The attempt made in *Early Law and Custom*, upon

* Pp. 90–92.

the basis of Ancestor-worship, to reconcile the several cases of Hindoo sonship with the Patriarchal Theory will, however, be discussed at length in a separate chapter.

As for the Slavonians, it is unnecessary to do more than repeat what has been shown of them in previous chapters.* Their earlier history, so far as it is known to us, affords no evidence in favour of the Patriarchal Theory, and furnishes not a little evidence against it, and especially against Agnation. Sir Henry Maine's views about them, on the other hand, have been formed on contemporary evidence; and all the evidence that could be quoted for those views is of comparatively modern date. The Slavs have been good Christians for many centuries beyond the date of the oldest part of such evidence; and no inference as to the institutions of the Primeval Family can be made from what it discloses to us of their social condition. The constitution of their House-Communities and their family usage, nevertheless, in no way countenance—on the contrary, they decidedly discountenance—the Patriarchal Theory. And it is certain that the Slavs fully acknowledge relationships through women; and that they give women and their descendants rights of inheritance, though postponing them to sons. It is needless to do more than mention again the warning conveyed by the case of the Hebrews as to inference from a limitation upon the succession of women such as is found among the Slavs.

* Chapters vii. and viii.

The Hindoos and the Slavs once considered, the weightiest part of Sir Henry Maine's evidence as to the prevalence of Agnation has been dealt with. Indeed, of particular statements as to Agnation made in *Ancient Law* to support the general statement that it is discoverable almost everywhere, only two remain to be noticed; and Sir Henry Maine has fully disposed of both in his later work. It is said in *Ancient Law* (p. 151), of the rule of royal succession known as the Salic law, that it has certainly an agnatic origin—"being descended from the ancient German rule of succession to allodial property;" and the rule in the Customs of Normandy which excluded uterine brothers from succeeding to one another in property derived from their fathers is spoken of as "a strict deduction from the system of Agnation under which uterine brothers are no relations at all to one another." As to the latter, however, *Early Law and Custom* discloses to us,* that the rule in the Customs of Normandy did not exclude uterine brothers from succeeding each other in property derived from their common parent, their mother—so that, though connected only through the mother, they were admitted to be each other's relations, and could inherit from each other according to their relationship; and there was no "strict deduction from the system of Agnation," or rather there was no question of Agnation at all. And for the Salic law another origin is found in *Early Law and Custom.* Instead of it being of agnatic origin because "descended from the ancient

* Note on p. 151.

German rule of succession to allodial property," the later view of its history declares the notion that it could be derived from any law of succession to property to be utterly mistaken. It was, according to this view, when the question of succession to the throne came to be mooted in France, a doubtful point whether a claimant deriving his right through a female could succeed to the throne of France or not—there was an absence, that is, of any known rule for the case which leaned towards Agnation; and the French decision against the right of our Edward III., who claimed through a female, was determined by feelings which are explained conjecturally, and which, as so explained, arose out of a natural misapprehension as to earlier practice. Something has been said of this view of the origin of the Salic law already,* and it appears to be erroneous. That rule of royal succession may have had, and had so far as anybody knows, nothing to do with the land law of the ancient Germans; but it seems beyond question to have been, in the apprehension of those who discussed it and those who adopted it, founded upon a provision of the Salic Code as to the succession to Salic land. What concerns us here, however, is to point out that the view of its origin propounded in *Early Law and Custom* makes it unnecessary to say more of the earlier view of its origin which derived it from a land law said by implication to have been agnatic.

It appears then that, of particular statements made

* See Note to Chapter vii.

by Sir Henry Maine to show that Agnation is discoverable almost everywhere, there are none now remaining except those relating to the Hindoos and Slavonians. There are some general statements which have yet to be noticed. But, before noticing them—the ancient Germans having been mentioned—it is impossible not to have in mind what Tacitus has handed down to us as to the relationships of the ancient Germans. A reference to their rule of succession to allodial property is bewildering. Montesquieu, no doubt, believed he knew what it was, because he knew what the Frank rule was, and had a hypothesis—in which he firmly believed—that the laws of the Salian Franks had come from the ancient Germans. But Montesquieu had nothing more or less than conjecture for this. And it need scarcely be said that what Tacitus has told us as to succession among the Germans, though, like many similar statements of modern travellers, not definite enough to be of use*—is, so far as it is worth anything, unfavourable to Agnation. What Tacitus has told us of the relationships of the ancient Germans, on the other hand, is perfectly intelligible. What it involves does not admit of the shadow of a doubt.

"A sister's sons," he says, "are as highly esteemed by their uncle as by their father. Some, indeed, look on this as the closer and more sacred bond of blood, and, in taking hostages, prefer to choose them on this principle, holding that such hostages give a surer

* "Failing sons, the next in succession are brothers, paternal uncles, maternal uncles."—Tacitus, Germania, ch. xx.

hold on the individual as well as a broader hold on his house." *

This shows, to begin, that the ancient Germans were very far from being in the agnatic state of relationship. On the agnatic view, a child and his uncle on the mother's side would not have been each other's relations —whereas, among all Germans, the uncle on the mother's side was as close to a child as its own father; and among some, he was even closer, so that, in taking hostages, a man's sister's son was taken from him in preference to his own son, as being the nearer and dearer to him of the two. The sister's son was preferred as a hostage, too, as affording a broader hold on the *house* of the giver—from which it appears that the house (*domus*) included kindred through women.† May we not ask in passing whether it may not have done so among the ancient Slavs also?

But the question now occurs whether there is any form of kinship known in which such a relation between a man and his sister's child as Tacitus describes to us is found to subsist. And the answer is that it is regularly found in communities in which kinship is acknowledged only through women. What Tacitus has told us of uncle and nephew among the Germans is true of the

* Tacitus, Germania, ch. xx.

† From a subsequent passage (Germania, ch. xxi.) it appears that compensation for blood was paid to the whole *Domus*. It follows that kindred through women shared in the compensation for blood, and that they were under the obligations of the blood-feud. It should be said that the suggestion as to the composition of the ancient German *Domus* made above has come from Professor Robertson Smith.

Nairs of Malabar—among whom paternity is unknown or unheeded; and is true of peoples like those American Indians (of whom something has been said in a previous chapter) by whom, though paternity may not be very uncertain, kinship is traced through women only. Among such peoples (where there is a father) the son is not counted of the same blood as his father, and is not his acknowledged relative—whereas he and his mother's brother are counted to be of the same blood, and are consequently each other's nearest male relations. The maternal uncle therefore is a child's natural protector among such peoples rather than the father; and it is not surprising to find him bestowing upon his nephew much of that affection which, where kinship through males has long been thoroughly established, people reserve for their own children.

Now it is not intelligible how such a relation between a man and his sister's son could arise where fathers had always been fully acknowledged. It is known that it exists where fathers are not fully acknowledged. And it is intelligible that, once formed, it should subsist, and even subsist long, after full recognition had been obtained for fatherhood. When we find it then, as Tacitus discloses it to us, among a people who acknowledge kinship through males, there seems to be room for only one conclusion about it. That is, that kinship through males was not always acknowledged among that people. And the statement of Tacitus appears therefore to contain convincing evidence that the Germans anciently acknowledged relationship through women only.

The ancient Germans then have to be classed with the Hebrews among peoples who, when they first become known to us, were clearly not agnatic. And, moreover, they have to be classed with the Hebrews among peoples who acknowledged kinship only through women before they acknowledged kinship through males. There appears to be no doubt, it may be said, that the early Greeks have to be placed with them in the former of these classes, and it seems difficult to escape the conclusion that they should stand, with Hebrews and Germans, in the second also.*

Of the more general, or less particular, statements as to the prevalence of Agnation which have been referred to as remaining to be noticed, one occurs in *Ancient Law* (p. 151), and is to the effect that the agnatic view of relationship pervades so much of the laws of the races who overran the Roman Empire as appear to have really formed part of their primitive usage. It would have been of the greatest interest and value to have had, for even one of the races alluded to, the native and ancient element in its laws sifted from the recent and extraneous in the manner suggested in this sentence, and the former shown to be pervaded by the agnatic view of relationship. But *Ancient Law* attempted nothing of this sort. And the sweeping generality it gave us instead seems to be effectually

* As to this, see "Kinship in Ancient Greece" in *Studies in Ancient History.* London: Quaritch, 1876. The succession law disclosed in the story of Meleager is the law peculiar to the system of kinship through women only. A man's heir, failing a brother, was his sister's son, and, conversely, the heir of a woman's son was her brother.

neutralised by other statements of a general sort which we have to notice, and which are to be found in *Early Law and Custom.*

It has already appeared from that work that the succession of a daughter's son, to her father, failing sons of his own, was "very widely diffused over the ancient world." This of course can mean nothing less than that the ancient world, very generally, was not under the influence of the agnatic view of relationship. But *Early Law and Custom* tells us further (p. 151) that there is "a strong probability" that "all sorts of ideas about succession to property" (and not agnatic ideas only) prevailed among the Aryan barbarians who overran Western and Southern Europe; and (the reason for that statement) it has to admit (p. 149) that many of the greatest races of mankind, when they first appear to us, allowed rights of succession, more or less, to women and the descendants of women—so that they were not agnatic. *Early Law and Custom* here seems to go directly counter to the general statement in *Ancient Law.* And so no more need be said of it.

Having now gone through all the evidence bearing upon the prevalence of Agnation which Sir Henry Maine has ever spoken of, we find that, after deducting what has in one way or another been given up, there remains to prove that Agnation is discoverable almost everywhere—in addition to the prevalence of Agnation at Rome of course—(1) the chance of Sir Henry Maine making something of the Hindoos—notwithstanding that the Hindoos, when we first come upon them, were certainly not agnatic (as to which, in a future chapter); (2) such evidence for Agnation as can be got from the social condition of Nineteenth Century Slavonians—who are not, and, it must be admitted, could not possibly be, agnatic; and, along with these, (3) distinct statements that the succession of daughters' sons was very widely diffused over the ancient world, and that many of the greatest races of mankind, when they first appear to us, allowed some rights of succession to women and the descendants of women—so that relationship with the descendants of women must have been fully acknowledged among them.

Sir Henry Maine has, in a passage in *Early Law and Custom* (p. 149), given a modified form of his statement as to the prevalence of Agnation—or rather a variant form, for it does not seem really intended to assert less. "The greatest races of mankind"—this is the new statement—"when they first appear to us, show themselves at or near a stage of development in which relationship or kinship is reckoned exclusively through the males." "They are in this stage," he adds; or (this is said to show that two views can be taken of the ancient facts which are unfavourable to Agnation) "they are tending to reach it; or they are retreating from it."

This new statement prompts the question, Which of the greatest races of mankind, when they first appear to us, were in the agnatic stage? Sir Henry Maine has not told us. And it is impossible not to see that he has never adduced a single real instance of Agnation.

Sir Henry Maine has not even offered any argument to show that something of common occurrence which is not itself Agnation—say a preference of males over females in succession, which clearly is not itself Agnation—implies a former prevalence of Agnation. Had he done so, no doubt, the argument for the Patriarchal Theory would have taken a rather complicated form. Patria Potestas, having disappeared, would have had to be inferred from Agnation; and Agnation, having also disappeared, would have had in its turn to be inferred from something else. He has trusted entirely to the frequency of Agnation. And yet he has never mentioned one clear case of Agnation other than the Roman case.

If what has been advanced in this chapter as to the Hebrews and others be well founded, such an argument as has just been spoken of could not succeed. Apart from this, it seems reasonable to hold that that which is very rare is exceptional, rather than to assume that it is normal, and the original of radically different forms which are actually met with.

CHAPTER XV.

AGNATION—CONCLUDED.

In the preceding chapters, in which Agnation has been treated of, it has been maintained that Sir Henry Maine has not enabled himself to recover lost cases of Patria Potestas by means of Agnation. An attempt has been made to show that his account of Agnation does not explain how Patria Potestas can have produced, or have been the foundation of, Agnation; to show that all we know of Patria Potestas discountenances the supposition that it can have had the effect attributed to it; and to show that, even on Sir Henry Maine's own statement, Patria Potestas could not have originated Agnation, or have been the foundation of it, unless people had systematically married outside the circle of their acknowledged relatives—so that this system of marrying, which is nothing other than exogamy, appeared, on his own statement, to be the real determinant of Agnation. These objections to his account of Agnation, of course, involve that it is a hypothesis which must be rejected, and, indeed, that, on his own statement, it is manifestly a bad hypothesis. It has further been maintained against this theory of Agnation that it so connects Patria Potestas and Agnation that it would be necessary, were it a hypothesis which otherwise seemed sufficient, to show how Agnation could survive Patria Potestas, and be found where no trace or memory of Patria Potestas was remaining—while this has not been attempted; and that therefore the occurrence of cases of Agnation, independently of Patria Potestas, were they forthcoming, would only tend to discredit and disprove the hypothesis. It has been pointed out that, while Sir Henry Maine's account of Agnation assumes that Agnation could survive Patria Potestas indefinitely, at Rome Patria Potestas survived Agnation.

Of course, a hypothesis, even if in appearance sufficient, cannot be received for more than a hypothesis—much less can it rightly be the subject of confident belief—unless it can be in some way sufficiently verified. And Sir Henry Maine's hypothesis, whatever may be thought of it, has been in no way verified. To make us believe that Patria Potestas and Agnation always occurred together in the early history of man he can only show us that they occurred together once, and that in a community very far from primitive.

Of the theory of Agnation propounded in this work all that need now be said is that the main points of it are deducible from Sir Henry Maine's attempt to show that Patria Potestas was the foundation of Agnation.

If what has been urged against Sir Henry Maine's account of Agnation—or even a part of it—be thought maintainable, then, were cases of Agnation ever so plentiful, Sir Henry Maine would be left so far as they were concerned, with one case of Patria Potestas found, so to speak, in a middle-aged community, by means of which to convince us that Patria Potestas was a universal institution among primeval men.

But when we go on to inquire as to the prevalence of Agnation, it appears that, so far as Sir Henry Maine has the means of showing us, Agnation is just as rare as Patria Potestas itself. If we are to believe in the universal prevalence of Agnation we must do so, it would seem, upon inference—actual cases of Agnation being, at any rate, not produced; and as Sir Henry Maine has never offered any argument to connect Agnation with forms of kinship which are found but which are not Agnation—since they show beyond all question the fullest admission of relationship for the descendants of women—we are left, so far as he is concerned, to make the inference without any reason. It would have been simpler, and quite as legitimate, for him to have inferred Patria Potestas from the existence of fathers, and then—on the strength of the theory connecting the two—to have inferred Agnation from Patria Potestas.

What we know of the Hebrews, however, to say nothing of other peoples, seems to make it clear that there can be no inference of Agnation in the cases upon which a belief in its prevalence would have to be formed. Granted that there have been cases in which Agnation existed or broke down, admission of relationships through women following, we appear to be even without means of recognising such cases—without means of distinguishing them from cases of which the precedent history was different. They would be in appearance identical with the case of the Hebrews, in which it is plain there was no Agnation.

And what is there to make us think they have been universal? What to make us think that they ever were numerous? What to make us think that, where they did occur, they dated from the beginnings of human history?

The sum of the matter so far is that Sir Henry Maine has got a single clear instance of Agnation whereby to make it probable that Agnation has prevailed everywhere; and with it a single instance of Patria Potestas—both occurring together in a community by no means barbarous, and neither of them short-lived in that community—whereby to show that Patria Potestas and Agnation were universal in the primeval family. And on the other hand, the researches into which we have been led in examining his statements, have disclosed to us that the early Hebrews and other races, when they first appear to us, were certainly without Agnation; while they show almost unmistakable signs of having acknowledged kinship through women only before they acknowledged kinship through men.

The instances which came naturally in our way in the preceding chapter, of ancient nations—Hebrews, Germans, Greeks—whose customs exclude the notion of their having been agnatic, are, of themselves, enough even to raise a strong probability that the kinship actually found among other early nations was, in general, not preceded by Agnation—whereas, from the occurrence of Agnation at Rome, we cannot get the inference that Agnation was primeval even among Roman tribes, or get any light as to the extent of its prevalence elsewhere.

And were all the instances in which actual Agnation has been found by observers, or even suspected, added to the Roman instance, the amount of reason there would be for thinking that Agnation commonly preceded among ancient nations states of kinship that were not Agnation would not be appreciably increased—there would still, that is, be none at all. There are, as has been said already, exceedingly few known cases of Agnation, and they all occur among non-Aryan peoples who may without rashness be suspected of having made a progress from the system of kinship through females only.

On the other hand, there are so very few known cases of Agnation that, Agnation appearing to be a stable system, and having shown good powers of endurance at Rome, we seem to be forced to conclude that it can never have been frequent, that it occurred in exceptional cases only, and therefore, that it cannot have been anywhere a primary form of kinship.

That the gentile bond, in those cases in which it became, with kinship through males, the limit of acknowledged relationship, appears to have been a factor in producing Agnation is, indeed, the best reason there is for believing that Agnation, though so little known to us, must have occurred to some extent. It appears to be a reason for believing this, at any rate, if we hold that kinship through males has, in general, been arrived at by a progress from a previous system of kinship through females only. Of the slight essay to describe the conditions of such a progress made in this work, a great part has been taken up with inquiry as to the circumstances which would favour Agnation, and there appeared reason to think that more or less of Agnation was to be looked for. It may be added that Agnation was, on the view submitted, likely to have occurred more rarely than a paternal power resembling Patria Potestas in its degree. The abandonment by a woman's relatives of her and her children to the man to whom they gave her in marriage, without which, it seemed, there would be no possibility of Agnation, might have sufficed for the establishment, for a time at least, of a high paternal power—for it would leave the man, so far as the woman's kindred were concerned, uncontrolled master of his family; while other conditions besides this seemed to be necessary for Agnation. Where, on the other hand, wife and children were not thus abandoned to the husband—where they were under the protection of the woman's relatives—on the view submitted, neither Agnation nor anything like uncontrolled paternal power were to be looked for; and it is consonant with this that we learn from Tacitus * that the German father was required to respect the lives of his children.

* Germania, ch. xix.

What has been said in this chapter is, of course, said subject to the result of the inquiry as to sonship among the Hindoos, which has now to be undertaken.

CHAPTER XVI.

SONSHIP AMONG THE HINDOOS.

In a previous chapter* we had occasion to describe the Levirate as it existed among the Hindoos and among the Jews, and to discuss the conditions on which it could be believed that the Levirate was derived from Tibetan polyandry. Referring the reader to what has been said in that chapter, it seems sufficient for our present purpose to repeat that when, among the early Hindoos, a man died leaving a widow and no children, it was the duty of his surviving brother to beget a son for him upon the widow; and that the child was counted in every respect equivalent to a legitimate son of the deceased, and continued his family, and was his heir—excluding his actual father from inheriting. Failing a brother, a kinsman, the nearest available, might fill what was normally the brother's part. It may be added that the Brahminical writers on Hindoo custom, regarding the Levirate with disgust and loathing, insisted that the connection between the widow

* Chapter xi. Page 156 *et seq.*

and her husband's brother ought not to subsist after one son, or two at most, had been born; and that they attempted in other ways to put what they thought a filthy practice under regulation. We do not know to what extent their injunctions were deferred to.

We are now to consider, among other matters, whether the Levirate has been accounted for independently of polyandry.

There is still, it may be said, a good deal of polyandry in India, and it is found both in the higher Tibetan form, in which the co-husbands are brothers, and in the lower form, in which the co-husbands are not brothers, of which indeed the typical example is furnished by the Nairs of Malabar. Nor is the British polyandry unknown. The evidence for Tibetan polyandry among the early Aryans, while small in quantity is excellent in quality—indeed convincing; not the less that it may be said it shows us polyandry not as a prevailing practice but surviving as an ancient custom amid a prevailing practice of monandry. Of the more direct and striking portions of that evidence a brief account will be found in an article on "The Levirate and Polyandry," which has already been referred to, or in Mr. J. D. Mayne's work on Hindu Law and Usage.*

It is evident, at any rate, that, in point of grossness, there is nothing to choose between the Levirate and polyandry. And the Hindoo family presents other

* "The Levirate and Polyandry." *Fortnightly Review*, 1877. *Hindu Law and Usage*. Madras and London, 1877.

facts which appear to indicate practices as bad as either or worse.

The list of sons enumerated by the early law writers, besides the legitimate son and the son of an appointed daughter—a daughter appointed to give an heir to her father because he had no son—includes the following: (1) the son begotten on the wife (by a man appointed for the purpose by the husband); (2) the son born secretly (as in the husband's absence); (3) the damsel's son (born before marriage, and who, if the woman did not marry, was among her father's heirs); (4) the son taken with the bride (unborn at the time of marriage); and (5) the son of a twice-married woman (not to be described briefly, and of whom it suffices to say that he also might be the child of a stranger). This discloses a view of the matrimonial relation which might be thought to show affinity to the Nair rather than to the Tibetan polyandry.

Sir Henry Maine has, in *Early Law and Custom*, offered an explanation of the Levirate and also of the peculiar forms of sonship just mentioned as having been acknowledged among the Hindoos. Mr. J. D. Mayne, a thorough-going supporter of his theories, had previously propounded an explanation of the same facts. As the Levirate is crucial for the Patriarchal Theory, it is proper that the views of both writers about it should be noticed here; and Mr. Mayne's view, as being the earlier and the easier to deal with, may be taken first.

Mr. Mayne* thinks he has accounted for the Levi-

* See his *Hindu Law and Usage*, pp. 48 *et seq.*

rate among the Hindoos independently of polyandry, by saying that, with them, it was an extension of the Niyoga—the appointment by a husband in his lifetime of a man to beget a son for him.

From a consideration of the facts relating to sonship among the Hindoos which have been mentioned, he deduces that, in the theory of the law, or rather on the old Hindoo way of regarding the matter, physical paternity was not material, was indeed of no importance, in connection with sonship; and that a man counted his wife's son as his son, whoever the father might be, because the wife belonged to him.* On this view, the Niyoga child, like all other children, was counted the husband's child because it was the child of his wife and he was owner of her; and it ranked high among children (this it certainly did), as having been begotten by a man selected by the husband. Ultimately the man proper to be appointed was a brother or, failing a brother, a near kinsman; but in all the early examples recorded, as Mr. Mayne points out, he was a stranger—so that there is no doubt that in early times a man might get a child through a stranger.

When Mr. Mayne goes on to show how, among the Hindoos, the Levirate (it may be so called for distinction's sake) was an extension of the Niyoga, that

* The only case (apart from the Levirate) in which he has found some difficulty in making this view square with the fact is that which, for a perfectly sound theory, might have been expected to be the easiest case of all—that of the son of an appointed daughter, in which a man who has no son takes as a son his daughter's eldest son.

there was in this case no husband living either to make an appointment or to own the wife and child is felt by him to be a difficulty. And so it is. The Levirate did not depend, as the practice from which he seeks to derive it did, upon a husband's will; and, with the Levirate, the affiliation of the child could not result, as with that practice it might do, from actual ownership of the mother. From Mr. Mayne's point of view, therefore, the Levirate plainly is not in the same position with the Niyoga. And Mr. Mayne perceives this; and perceives, too, that it cannot be brought under the theory of sonship of which he found in the Niyoga an apt enough illustration. Indeed, in trying to show it to be an extension of the Niyoga, he gives up his theory of sonship. He says that, with the Levirate, "the element of fiction was introduced," and he throws out the suggestion that the fiction may have been that the husband was regarded as surviving in the wife—which, whether a fiction likely to be resorted to or not, would be a fiction consistent with his theory. But, merely mentioning this, so far as he commits himself to any meaning, his view is that there was introduced with the Levirate a fiction, or rather a false pretence, of *paternity*—of the paternity, that is, of the putative father—that one thing which, in the Niyoga, was treated as absolutely of no importance.

Mr. Mayne does not attempt to show how, in extending the Niyoga, a fiction so inconsistent with it came to be used. And this, of course, means that he has failed, on his own view of the Niyoga, to account for the Levirate as an extension of the Niyoga. The Levirate child could not be reckoned a son of his putative father in virtue of those facts which, on Mr. Mayne's view, gave the Niyoga child the position of a son; and (according to Mr. Mayne) he was so reckoned in virtue of a fiction which is not derivable from the Niyoga. If there was a fiction of paternity with the Levirate, Mr. Mayne leaves us to seek for ourselves an explanation of it—and therefore leaves it as open a question as ever whether it was derived from Tibetan polyandry or not.

All this, however, must seem very like trifling; seeing that, beyond the inquiry whether the Levirate was an extension of the Niyoga, there is the question, what originated the Niyoga itself? The early cases in which the person called in was a stranger clearly do not belong to a regulated practice, and appointment is not the word to apply to them—husbands could do what they pleased, and it is more accurate to say that they selected or requested than that they appointed. When such cases could occur, there can have been no regulated practice—though what afterwards was the regulated practice may have been customary practice even then. At any rate, in the Niyoga as embodied in the law, we find the choice of appointment restricted exactly as it was in the Levirate; and this raises in connection with it the very question raised by the Levirate. Strange as it may seem, Mr. Mayne has not noticed this. The element in the Levirate which can be taken as indicative of Tibetan polyandry is the substitution for the husband of his brother, or, brothers failing, of his kinsman. The same substitution occurred in the Niyoga—and Mr. Mayne has not given this a thought. If we grant to him, however, that every child of a married woman was counted her husband's child, it is quite clear that there remains the question, why was it required with the Niyoga (as with the Levirate), that the father should be the husband's brother or kinsman? To overlook this question is to miss what is the problem of both Levirate and Niyoga altogether.

Mr. Mayne has, moreover, not thought of considering why, among the early Hindoos, physical paternity was, as regards paternal right, so immaterial as he takes it (justly, we think) to have been, or what may be involved in that. He appears to have taken this to be absolutely unimportant. Having found (or thought he had found) among the early Hindoos wives owned, paternity of no account, and maternity so far the determinant of sonship, he concluded forthwith that he had got to the primitive state, and had found it quite consistent with the Patriarchal Theory.

It remains to be added that Mr. Mayne has tried to account for the Levirate independently of polyandry in the Hindoo case only, and by circumstances which, so far as he discloses, are peculiar to that case. He has not troubled himself about other cases. But the Levirate has been widely spread; and all cases of it, whatever minor differences there may be between them, agree at least in this, that a man "raises up seed" to his brother or kinsman deceased. Is it likely that men were led to that in one place by one set of circumstances, in another place by another set—and that there may be as many explanations of it, therefore, as there are instances of it? On the contrary, are we not forced, or almost forced, to think that its origin was everywhere the same; that what would be a good account of it in one case would be a good account of it in every case; and that an account of it evolved from one case which does not suffice for the others must be bad? If this be conceded, Mr. Mayne's view, could it be entertained, would involve rather startling consequences. Had the early Hebrews the Niyoga, and were they as indifferent as the early Hindoos who were the fathers of their children? Perhaps so; but this has never been laid to their charge; and it imputes to them practices in comparison with which Tibetan polyandry may be deemed respectable.*

* We do find among the early Hebrews a practice analogous to the Niyoga, but one which indicates a different state of social relations from that which is being considered among the Hindoos—that *a wife* could appoint her female slave to bear a child for her, which should be counted her child.

Upon this Professor Robertson Smith contributes the following note, which, besides its bearing upon the matter in hand, contains additional Scriptural evidence of weight against the Patriarchal Theory.

"The slave in this case," he says, "is the slave of the wife, and does not cease to be so, Gen. xvi. 6, 9 (a chapter which is by a different and older hand from Gen. xxi., in which, at v. 12, Hagar is Abraham's bondwoman). It appears, then, that the case would most naturally arise in a state of society where the child belonged to the mother rather than to the father, and of this we have a striking piece of

If it be not conceded that there must be a common explanation for all cases of the Levirate, a theory restricted like Mr. Mayne's to a single case could not, even if accepted, appreciably lessen the difficulty which the Levirate makes for the Patriarchal Theory.

While, as we have seen, a careful study of Indian facts has impressed upon Mr. Mayne—albeit a supporter of the Patriarchal Theory—that physical paternity was, in the theory of ancient Hindoo law, not material to sonship, Sir Henry Maine's account of the Levirate sets out from a view of an entirely different sort. Putting aside the law-books (though it is to such writings that Comparative Jurisprudence makes its appeal), making no appeal to the Hindoo writings that are more ancient than them, Sir Henry Maine *assumes* the Patriarchal Theory, and makes his start from that.*

evidence in 1 Sam. i., where the mother has power to give her child out of her family by devoting, or, as she calls it, "giving" or "lending" him to Jehovah, so that he leaves his own people and is incorporated among the ministers of the Sanctuary. By the later law (Num. xxx. 3–9) the vow of a daughter or a wife is not valid if her husband disallows it; but even if this held in the time of Samuel, and Elkanah had to confirm his wife's vow, it is clear that she could not make a vow of a son that was not *hers* but the father's. Note, by the way, that as only daughters in the father's house, and not sons, require the consent of the father to their vows, Num. xxx. proves that sons were not *in manu*."

* In his *Early Law and Custom*, London, Murray, 1883. This assumption cannot be fully treated of until the Hindoo facts have been set forth. But something may be said of it at once. Sir Henry Maine begins, then, with some observations on Ancestor-Worship—a large and very difficult subject, of which he does not profess to have made a special study. He concludes (at p. 75) that, when ancestor-

This assumption gives him, for the continuation of the Hindoo family, to begin, a basis of legitimate sons. His theory always involved that these could be added to through the fiction of adoption; and finding that the ancient Hindoo reckoned as his sons the (as we should say) illegitimate son of his wife born before or after marriage and the illegitimate son of his daughter, he suggests that such persons also were assumed as sons in virtue of some fiction, and dubs them fictitious sons. A man who had a daughter, and no son, could "appoint" his daughter to bear a son for him; but, instead of throwing doubt on the Patriarchal Theory, this only appears to him extremely interesting as probably marking one of the points at which the right of women to inherit made its way into the strict

worship arose, paternity was fully recognised. He proceeds (at p. 86) to a vastly stronger proposition—that, for an explanation of ancestor-worship and its legal consequences, we must assume that when ancestor-worship arose, "the Father of each family appeared to them in the form in which he constantly shows himself on the threshold of jurisprudence"—that is, with the position and powers of a Roman Paterfamilias. Here, as a reason for the assumption of the Paterfamilias and the Patriarchal Theory, is the statement which reiteration has made so familiar. But we have seen in what number of cases Sir Henry Maine can show us the Paterfamilias "on the threshold of jurisprudence." We had been led to look for him on the threshold of Hindoo jurisprudence. But it is just because he cannot find him there that Sir Henry has gone to Hindoo ancestor-worship in search of him. The reason given is a bad reason—not good in fact—and the assumption, not being justified by it, is the merest assumption.

The "legal consequences" to which Sir Henry Maine makes vague allusion are those very facts which impressed upon Mr. J. D. Mayne that paternity was not the determinant of sonship in the theory of Hindoo law.

agnatic system of kinship and succession which (he assumes to have) prevailed among the early Hindoos. When he comes to the Levirate, he begins by setting forth what the facts were. He tells us (*Early Law and Custom*, p. 100) that, "in the opinion of some of the Hindu doctors," a man who had died leaving no children, real or fictitious, might have an heir born to him by means of the Levirate; and that "here and there" (p. 102), the practice of substituted begetting received an extension even more revolting to modern delicacy—in the appointment by a husband of another man to beget a child for him. It thus appears that, instead of regarding the Levirate (so to call it still for convenience sake) as an extension of the Niyoga, as Mr. Mayne does, he regards the Niyoga as an extension of the Levirate. Both practices, however, being now exhibited as of paltry dimensions, he proceeds to account for them as follows:

"Let us suppose," he says, "that in a particular society an intense desire has arisen for male issue, whether through its worship of ancestors or otherwise. Let us assume that in a particular case actual issue of the father's loins is impossible. There are no daughters. The accepted fictions, by which sons are created for the sacrifice, cannot be made serviceable. What is to be done, that the name of the aged or dead man be not put out on earth nor his lot placed in jeopardy beyond the grave? Now all ancient opinion, religious or legal, is strongly influenced by analogies, and the child born through the Niyoga is very like a real son. Like a real son, he is born of the wife or the widow; and, though he has not in him the blood of the husband, he has in him the blood of the husband's race. The blood of the individual cannot be continued, but the blood of the household flows on. It seems to me very natural for an ancient authority on customary law to hold that under such circumstances the family was properly continued, and for a priest or sacerdotal lawyer to suppose that the funeral rites would be performed by the son of the widow or of the wife with a reasonable prospect of ensuring their object. The very differences of opinion which arose on the subject in the most ancient Brahmanical law-schools seem to me exactly those which would be provoked by a plausible and yet non-natural contrivance." *

The first sentence of this passage shows a desire to make the explanation apply to all cases of the Levirate. But it is plain that the passage has been written with Indian facts only in view.

The son born through the Niyoga, we are told in this passage, was "very like a real son." But it is scarcely necessary to point out that he could not appear so to a people in whose theory paternity was of the first importance. A married woman's child by another man than her husband does not to modern people seem "very like" a child by her husband. That the man is her husband's brother does not make it more like—that "the blood of the household flows on" in the child makes no difference. And the reason is that we think

* *Early Law and Custom*, pp. 106–7.

paternity as important as maternity. That being so, the blood of the household does not satisfy us any more than the blood of a stranger. And it could not have satisfied the Hindoos had they had the modern view about paternity. It seems to follow that they had a lower or less exigent view, consistently with which the child could pass for the son of the husband. How much lower then—how much different from our view? Sir Henry Maine has not assisted us here, as he might have been expected to do; seeing that it is clear that the Niyoga child could not to people of all views alike appear "very like a real son."

Fortunately, we are not dependent upon conjecture in the matter. The view on which the Niyoga was practised is clearly disclosed to us by Apastamba, and it most amply accounts for the practice. "They declare," says Apastamba, "that a bride is given to the family (of her husband and not to the husband alone)." * To people holding that view of marriage, a Niyoga child would seem not merely "very like" a real son, but quite a good son—it was enough for them that he was the child of the wife by one of the family—their theory did not exact that the husband should be the actual father. Need it be said that it is a view of marriage that might have come straight out of Tibetan polyandry?

Apastamba is by Sir Henry Maine taken to be one of the most ancient and one of the greatest of autho-

* Apastamba II., 10, 27, 3. *Sacred Books of the East*, Vol. II., p. 164. Oxford, 1879.

rities; but in this instance, and in another that will have to be referred to, he ignores Apastamba's evidence as to the views or ways of his countrymen. What Apastamba discloses, however, supersedes conjecture as to the view on which in his time the Niyoga was practised; and will explain perfectly its being an established custom. Sir Henry Maine, nevertheless, has felt free to set it aside and to conjecture for himself. And his conjecture is that we have in the Niyoga "a plausible and yet non-natural contrivance" for continuing *the family*.

He had to account for the Niyoga, however, not as a means of continuing *the family*—it was not that which was in danger of becoming extinct—but as a means of continuing the line of a particular member of the family. He has therefore not faced his problem. And he has really not even taken a step towards accounting for the Levirate independently of polyandrous views.

It must be said, moreover, that the whole view of Hindoo facts which culminates in this account of the Levirate appears to be untenable.

Sir Henry Maine, as we have said, has assumed—not by way of hypothesis—that the Hindoos started with the Patriarchal Family. He has accordingly had to treat the facts available for judging what their early family life really was as showing aberrations from this primitive model (because they are *ex facie* irreconcilable with it). It then became his task to account for such aberrations—no easy matter. His preliminary assumption seems to have here led him into underrating the importance of the supposed aberrations, and into assigning for them a time of origin which would allow something like a breathing space for itself. Facts which look as if they must be very early are accordingly presented to us as late. Facts which used to be thought important dwindle until it scarcely matters whether a good account is given of them or not. And other facts which should have been reckoned with have been neglected altogether.

What has now been said will, we venture to think, be justified in course of this chapter.

It might be added that Sir Henry Maine has, with the utmost confidence, treated as the oldest and most authoritative of Hindoo law writers certain authors who have generally been regarded as neither the oldest nor of the highest authority, to whom of late a great antiquity has been assigned upon reasoning the feeblest and most inconclusive—and which, on the best view of it, could only justify a tentative opinion. But this need not be urged. For those authors do not contribute to the establishment of the Patriarchal Theory. Apastamba is one of them. We have already had occasion to make use of him, and we shall have again to refer to him by-and-by.

We go on now to consider how Sir Henry Maine, upon his assumption of the Patriarchal Theory, accounts for the sonship of the illegitimate, or adulterine offspring whom Hindoo custom classed as sons. Instead of their being first so classed when the Hindoos were uncivilised and not too nice, he takes it that they came to be counted in some degree as sons only when the Hindoos had fallen off from their original strictness under the influence of religion, which made them earnestly desire to have male offspring. At times, as has been said already, he seems to think such persons were reckoned as sons in virtue of fictions comparable to the fiction employed in adoption, and he calls them fictitious sons. He has made no suggestion as to what the fictions were; and it is idle to speak of fictions unless the fictions can be indicated. But this need not be insisted upon; for he goes on to explain how, consistently with Patriarchal custom, such persons might have been counted as sons, and here he dispenses with fiction altogether.

"They are all," he says of the illegitimate and adulterine sons, "the offspring of women who are under the shelter of the household, or who are brought under it. These women are under the protection of its head; they belong to him, and the status of their children is settled by the well-known rule which, in Roman law, would settle the status of a slave." Paternal power and protective power, he proceeds, are inextricably blended; even the slave was in some sense a member of the family; and those children were "permitted to rank as in some remote degree sons" because their mothers were protected by the head of the household, and because they were themselves protected by him.* This is not so precisely expressed as could be desired; but

* *Early Law and Custom*, pp. 95–99.

the effect of the whole is that the child was ranked as "in some remote degree" a son because both his mother and himself were under the protection of the man who was counted "in some remote degree" his father. He ranked as a son, as far as he did so, because he was his mother's son and both were protected by the head of the household—who had what was equivalent to paternal power over him.

The use of a fiction, be it observed—at any rate, if it operated like the fiction of adoption—would have been more effective. Adoption would have made the child of the adopter's stock and family—a son in the fullest sense. The protection of mother and of child, however, according to Sir Henry Maine, made the child rank as a son only in some imperfect sense.

What he has here tried to show is something less than the case required. But that may be passed over for the moment. It is proper first to point out that he has not kept in view what being a son involved among the Hindoos. It at least involved that son and father should be of the same stock (gotra). And it is evident that mere protection could not introduce a man into a stock to which he did not belong. This explanation therefore is at any rate defective. And next, there is something more serious to say of it. The Hindoo facts appear to exclude it. There are several facts, indeed, which, taken singly, are enough to put it out of the field.

Considering the passage above referred to, one may perhaps discern an argumentative reason for defining a wife or a daughter—for the heirship of the child of an unmarried daughter is among the facts to be reckoned with—as "a woman protected by the head of the household." But it is a very odd definition, and very far from being exhaustive—it would apply to a female slave. It takes no account of the relation of husband and wife, of father and child, which, in the one case and in the other connected the man with the woman. And yet both of these relations generally involve consequences of importance. When a relationship to the man arose through the woman, surely it is necessary to inquire how far it was attributable to the relationship previously subsisting. It would seem that a theory which in such cases carefully excludes consideration of that relationship has been planned so as to give it every chance of being wrong.

Let us look at the case of an unmarried woman's child. So long as the woman remained unmarried the child belonged to her father, and was among his heirs. In virtue of what? Is it absurd to say that it was simply in virtue of the tie which connected the man and the woman? On the contrary, there is no need in this case even to consider whether Sir Henry Maine's suggestion can account for the child's position. There is plainly no room for it. For the relationship which was precedent to protection of the child and laid the foundation for it—the relation between father and daughter—accounts for the child's position. An unmarried woman was of the stock of her father, and he had a father's rights over her. Her child might have belonged to its own father. But by Hindoo custom it never did. It might have been of its father's stock. But by Hindoo custom it never was. The alternative was that it should "follow the mother" and belong to her father who had rights over her. And by Hindoo custom it was her father's—and that because its mother was his. Moreover, it was of his stock—which was also its mother's. That is shown by its having been his heir. Being his daughter's child, and of his own stock, it was among his heirs. Here we see that parental right over the mother was the foundation of parental right over her child. And parental right over the child was the basis of the protection given by the woman's father, and not in any way an effect of it.

That without adoption—for of that nothing is anywhere said—the child of an unmarried woman was of her father's stock, which was also his mother's, yields, it would seem, irresistibly the inference that the stock could be transmitted through the mother—that the child in this case took blood or stock connection from the mother; an inference to be noted, because of its relation to the familiar facts which first raised doubts about the Patriarchal Theory.

To proceed with this case—when the woman married, the Hindoos being exogamous, she married a man of another stock than her own, and she passed into his stock. The child perhaps could be retained in her father's family—and if the father had no son he would be retained as a matter of course. But if he were taken over with his mother—and that "the damsel's son" was the descriptive title of a class of sons seems to show that this happened commonly and in pursuance of a customary arrangement—what founded the connection between him and his mother's husband? Surely it was marriage—his mother's marriage. Protection came afterwards—as a consequence of the connection so founded. He was assigned to her husband along with her. That he was now counted her husband's son shows that he was now counted of her husband's stock; and that shows very clearly that he "followed the mother" in the change of stock. Mother and child passing (by admission to his sacred rites, no doubt) into the husband's stock, the man got a wife and a child at the same time—a thing not perhaps "patriarchal," but, nevertheless, by no means unexampled.

Here we find in marriage, as in the previous case we found in the relation of father and child—those elements in creating relationships which Sir Henry Maine has forgotten—the explanation which he has sought in mere protective power. The problems of this case, however—how to account for the child being first affiliated to the stock and family of the mother's father, and afterwards to the stock and family of her husband; and how to explain this happening so systematically as to give rise to a legal description of a class of sons—seem scarcely to have had full consideration from him.

As to the adulterine offspring of a married woman, while the law always recognised their sonship, and the most ancient custom gave them to the husband, it appears to be quite clear that the husband's right to them was founded upon marriage and nothing else.

It is in the salient or peculiar features of any body of custom—at the points in which it differs from other bodies of custom or law—in what is distinctive in the practices permitted by it—that we can discern the basis, if there be any, on which the whole has been built up. And if all that is distinctive is found to be consistent, and referable to the same basis, and that which is not distinctive opposes no difficulty, we cannot hesitate to believe that we have found the basis of the whole. Let us search then—the quest will not occupy us long—in the differentiating points of Hindoo practice for the basis of Hindoo family law.

We have seen that, in very early times, a man could appoint, or call in, a stranger to beget a child for him, and that the child was his own and not the actual father's. The child was born to be his son; and that by itself is enough to dispose in this case of Sir Henry Maine's protective power theory—and enough, indeed, to make an end of it altogether as an account of the putative sonship of the Hindoos. Moreover, it was not "in some remote degree" a son, but a highly-valued son; and it was not brought into existence for lack of children but because the qualities of the sire gave promise of a child of superior excellence. Next, it is plain that at this early period a child did not belong to the begetter, or at least that it need not belong to him. It is plain, too, that at this period a Hindoo need not have been the actual father of his own son. And since the child was certainly not the father's child in virtue of physical paternity, the foundation of his right to it can only have been that he was the husband of the mother. It was his in virtue of his marriage. It was his because it was the child of his wife.

The case, which has already been dwelt upon, of the son of an unmarried woman, carries us a step further. The child "followed the mother," and was of her father's family, the actual father having no right to him. Here there was nothing which gave the father of her child a right over the woman, and we see that the paternity gave him no right to the child—that is, that the child did not belong to the begetter. We find also that the child belonged to the woman's father, to whom she herself belonged—that is, that it belonged to him who possessed rights over the mother. Bearing in mind that, when a woman married, the rights of her father were transferred to her husband, and recurring to the case last dealt with, it becomes quite clear that the appointed person could not in that case, as being actual father (even had there not been an implied contract to the contrary), have any right to his own child; and also that the rights over the mother which marriage had passed from her father to her husband were enough to explain how the child was counted the husband's. The child was his because he alone possessed rights over the mother.

The case of an appointed daughter carries us further still. In that case, when the woman married and her father had no son, the right to her first son was reserved by her father. Hence we see that, though the husband might be, and in all probability was, the father, that fact gave him no right to his child by his own wife. The son did not belong to the begetter—even with marriage. It did not even belong to the husband of the mother. It went back to the mother's father. Why? Because the father, who had given the woman to the husband, had not given her unconditionally, but with reservation of one of her offspring to himself. That which he could reserve must have been part of that which in ordinary cases was given—he could reserve nothing else. And we see that he could reserve one of his daughter's offspring even when it was also the offspring of her husband. We know that a woman's offspring, while she was unmarried, belonged to her father. It follows that a husband's right even to his own children by his wife sprung out of the right which at marriage passed to him from her father, and that he got no right to them merely as being actually their father.* And while he got the

* Apastamba, the only Hindoo writer who gives full logical effect to the doctrine that a son belongs to the begetter, consistently omits mention of the appointment of a daughter. He prefers to allow to a daughter rights of inheritance, failing sons. It will be seen hereafter that, in giving his cherished dogma full play, he has had either to forbid or to omit everything that was distinctive in Hindoo family custom. But he stands alone among early authors; and his work, so far as it relates to family custom, is throughout a protest against the existing customs and not an account of them. This appears beyond doubt from disclosures made by himself. It may be mentioned that while, as stated above, he admits daughters to inheritance, he does not recognise adoption. The Patriarchal Family would not have pleased him.

right to have children by his wife who should be his own children, if he chose it, he got more. The first case that has been noticed is enough to show that anciently Hindoos chose to get more.

How far the right passed to the husband at marriage might extend is—to go on to another case—to be gathered from the Levirate; from which we see that a man might have issue of his wife counted his even after he was dead. Children so counted to him were children by his brother. But it was not any children of his brother who could be thus counted to the deceased. It was necessary that they should also be children of the woman to whose issue he had acquired right by marriage. The Levirate shows us once more that physical paternity by itself gave no right, and that Hindoo fatherhood was not dependent upon it.

It may be said by the way that the facts considered above seem to point beyond serious question to marriage among the Hindoos having, in early times, commonly proceeded upon contract—and that, in the ordinary case, a contract for the transference from father to husband of a woman and with her, her issue. A right—a substantial, a highly valued one—passed from the father to the husband. On occasion, a part of it could be reserved. Out of what could the custom of reserving this part come if not out of a system of bargaining? The Hindoos were exogamous, and among early exogamous peoples, who had got beyond capture, purchase was the only way of getting a wife. A man who had to look for his wife to people who, being strangers to him in blood, owed him no duty, could get a wife (putting capture aside) in no other way.

The Hindoos retained in historic times a remainder of marriage by capture. Marriage by the Asura-rite—that is, by purchase, though not approved of by the authorities, was always recognised as among the forms of marriage. Moreover, it was a custom, coming down from very early times that, at the time of marriage, the bridegroom should make a gift—and a very substantial one—to the bride's father. This it was the duty of the latter to return. But why was the gift made? How arose the custom of making it? Was there not a time when the father did not return it? Apastamba protests that the use of the word sale, in reference to these transactions (which occurs in some Smritis), is metaphorical. But, plainly, we have here a symbolism of purchase; and the word "sale" would have literally described that which was symbolised.* We need not hesitate then to conclude that marriage was commonly among early Hindoos, as their affiliations of children indicate it to have been, an affair of sale and purchase. And it may further be concluded—since one

* Apastamba II., 6, 13, 12. The passage is as follows: "It is declared in the Veda that at the time of the marriage a gift for (the fulfilment of) his wishes should be made (by the bridegroom) to the father of the bride, in order to fulfil the law. 'Therefore he should give a hundred (cows) besides a chariot; that (gift) he should make bootless (by returning it to the giver).' In reference to those (marriage-rites) the word 'sale' (which occurs in some Smritis is only used as) a metaphorical expression, for the union (of the husband and wife) is effected through the law."

child could be reserved—that the contract did not in early times *of necessity*—whether commonly it did so or not—carry to the husband the right to the children.

It may be added that Hindoo custom always allowed a husband to communicate to another man by contract the right to have a son by his wife. The "son of two fathers" was the child of a married woman by a stranger, to whom her husband had made her over upon contract—for a price—to let him have a son by her. Where such a contract was made, the child might, if the contract were to that effect, belong entirely to his actual father—of course, in virtue of the contract. But, in general, he was to be the son of both, and so was called the son of two fathers. The husband of the mother, though not his father, had him as a son in virtue of his right to the woman's issue. The actual father had him as a son in virtue of the contract. He could perform the usual rites for both his fathers. But we learn from Gautama that the actual father's right was rather precarious. For he might lose it by leaving the rearing of the child to the husband. The case is curiously illustrative of the ideas connected with Hindoo family right.*

Having considered the salient or differentiating features of Hindoo practice, we have now found that, by Hindoo custom, (1) the son certainly did not belong

* The place of contract in connection with early marriage is one of many facts which show that the acceptance of the famous maxim that the movement of progressive societies has been a movement from status to contract will have to be reconsidered. As regards marriage and the family it is clear there has been a movement from contract to status.

to the begetter (it has seemed necessary to show this clearly—though we may be sure the doctrine negatived has never been embodied in the usage of any people); (2) that a husband even did not acquire right to his wife's child in virtue of his being actually its father; but that (3) that was included in the right which passed to him from his wife's father at marriage; and (4) that, in the ordinary case, the right then passed extended to all the woman's issue, whether actual issue of the husband or not. There is nothing inconsistent with these deductions in any part of Hindoo family custom.

It is in full harmony with them that adulterine children and the child "taken with the bride" were counted among sons. They are so counted by every old Hindoo law-book except that of Apastamba—which omits them. But convincing evidence as to the sonship of adulterine children is to be got even from Apastamba.

As matter of theory, or rather of metaphysic—the general practice (that is, as we should say, the actual law) unaffected—the sonship of adulterine children seems to have been more or less in controversy among speculative Hindoos from an early period. The opinion hostile to it was the dogma that the son belonged to the begetter; and the ingenuity of its supporters put it in a form in which, its antagonism to custom being admitted, it was not unlikely to have much influence upon practice. The ways of this world being against them, they appealed to the ways of the next. They represented that a son (though belonging to the husband of the mother in this world) belonged to him who had begotten him in the world of Yama—so that no spiritual benefit was to be got by the reputed father through him. And we learn this very clearly from Apastamba.

"They quote also," says Apastamba "(the following Gâthâ from the Veda): '(Having considered myself) formerly a father, I shall not now allow (any longer) my wives (to be approached by other men), since they have declared that a son belongs to the begetter in the world of Yama. The giver of the seed carries off the son after death in Yama's world, therefore they guard their wives, fearing the seed of strangers. Carefully watch over (the procreation of) your children, lest stranger seed be sown on your soil. In the next world the son belongs to the begetter, an (imprudent) husband makes the (begetting of) children vain (for himself).'" *

Here we find a careless husband, who had been content with the children his wives gave him, because he regarded their children as his own, alarmed at the suggestion—represented as new to him—that they would not be his in the world of Yama. That they could belong to anybody else in this world the new doctrine did not suggest. It threatened careless Hindoos with what would happen in the world of Yama to make them cease from relying upon what was settled among the people they were living with.

* Apastamba II., 6, 13, 7.

If proof were needed, there could not be clearer proof that a child did not belong to the begetter in this world. There seems to be no doubt entertained, says Mr. J. D. Mayne, commenting upon this passage, that the son begotten on a married woman by a stranger belonged in this world to her husband.

It is Hindoo practice we are concerned with, and especially ancient practice. And here we have it disclosed by Apastamba—himself an extreme supporter of the reforming doctrine—that, in ancient times at any rate, Hindoo practice did not give the son to the begetter but to the husband of the mother.

Apastamba has given effect so thoroughly to the doctrine that the son belongs to the begetter that he has either omitted or forbidden everything that was distinctive in Hindoo custom—discarding among other things, that sonship by adoption of which the Patriarchal Theory has made lavish use. But he stands quite alone. And as we learn from himself—besides having other conclusive evidence—that ancient practice was against his view, and as his authority (if he really was a very early writer) did not affect subsequent practice, his opinion, however interesting, seems only to prove that there were Hindoos in his time (whatever it may have been) who strongly disliked the ways of their countrymen. He does not at any point say a single word which can suggest that he was making a stand against corruptions of recent introduction. On the contrary (as will be seen from the passage quoted on the next page) he admits that precedents the most ancient, and seemingly the most weighty, could be quoted against him—besides disclosing incidentally that the general practice of antiquity was against him. A thorough-going dogmatist, he laid down the law as, on his dogma, it ought to be. And it was with spiritual consequences he threatened those who did not conform to the law so laid down.*

* See Apastamba II., 6, 13, 6–11. The full text is as follows: "A Brâhmana (says) 'The son belongs to the begetter.' Now they quote also the following Gâthâ from the Veda: '(Having considered myself) formerly a father, I shall not now allow (any longer) my wives (to be approached by other men), since they have declared that a son belongs to the begetter in the world of Yama. The giver of the seed carries off the son after death in Yama's world; therefore they guard their wives, fearing the seed of strangers. Carefully watch over (the procreation of) your children, lest stranger seed be sown on your soil. In the next world the son belongs to the begetter, an (imprudent) husband makes the (begetting of) children vain (for himself).' Transgression of the law and violence are found among the ancient (sages). They committed no sin on account of the greatness of their lustre. A man of later times who seeing their (deeds) follows them, falls. The gift (or acceptance of a child) and the right to sell (or buy) a child are not recognised."

The fact stated about the ancient sages is what is important. It would be mere silliness to ask us to take the explanation along with it. A Hindoo jurist had to maintain that the law had been from the beginning as he was declaring it. And he had a simple way of accounting for the sages having conformed to a ruder sort of law.

Gautama, more practical than Apastamba, though of the same school, lays down that "a child belongs to him who begat it, except if an agreement (to the contrary has been made)." (Gautama, 18, 9, 10). Apastamba admits no qualification of his principle that "a son belongs to the begetter"—and rightly. To qualify a view of that sort is to give it up. But Gautama wished to state actual law, and to do that and not to qualify his principle was impossible. It is plain too that he desired to put on a thoroughly good basis the position of a Niyoga child. Notwithstanding his general rule, he includes

That so high (and in truth, for this world, so impracticable) a view about paternity should have been broached among the Hindoos, doubtless at a rather early time, seems to show that, among metaphysical Hindoos, from an early time thinking had been going on about the connection between parent and child. But customs are not made by metaphysicians—though metaphysicians may try to amend them*—nor are they based on metaphysical theories. It appears from Manu, too,† that the metaphysicians were very far from being agreed among themselves. Manu, though holding the connection of father and child superior to the connection of mother and child, distinctly and strongly sustains the established usage which assigned to the husband as children all offspring of his wife. He held that the man who had rights over the woman was her "lord"; and was entitled to her child by a stranger, just as the owner of land in which seed had been sown by a stranger without agreement, would be entitled to the resulting crop—and apparently for the same reason, because the receptacle, as he says, is more important than the seed.‡

the classes of adulterine and illegitimate sons in the family, and gives one of them—the "son born secretly"—equal rights of inheritance with a legitimate son. The others he admits to inheritance only on the failure of preferable heirs—and then only in a fourth of the estate.

* Were views like Apastamba's most likely to occur where practice did not countenance them? We find them among the Greeks as well as among the Hindoos. At Rome, where there was Agnation, thinking broke it down.

† Manu ix. 32; x. 70.

‡ Manu ix. 48–55.

Writers on the laws learned to discriminate between the various sorts of persons who were by usage regarded as sons, to arrange them in classes, and to make or suggest some distinctions as to their relative rights. But they had, in deference to usage, to count them all as sons. And their writings show that two principal views had been mooted as to the foundation of paternal right—views which of course came into actual conflict only over those cases in which, while custom gave sonship, actual paternity was admittedly wanting; one of these views being intended to sustain and account for actual usage, while the other was a protest against usage, and intended to revolutionise it.

It is perhaps scarcely worth while to add that the family of the Patriarchal Theory would not have satisfied Apastamba. But the use of adoption is as inconsistent with the dogma that the son belongs to the begetter as the sonship of adulterine children, of Niyoga children, of Levirate children, or any other distinctive feature of Hindoo family law; and accordingly he has not recognised it.*

The Hindoo customs have come down to us from an extremely remote period, while the Irish laws took the shape in which we have them at a time comparatively recent, and certainly have not escaped Christian influence.

* Sir Henry Maine has himself noticed (*Early Law and Custom*, pp. 97–8) that, among early Hindoos, the place of adoption was not an important one. It afterwards became a very important one. It does not seem to have occurred to him that this, so far as it goes, makes against his theory—which supposes that free use was made of adoption in the primitive family.

Nevertheless, as to some of the matters now being dealt with, there is a surprising amount of coincidence between the two—enough to show that the more modern has been arrived at by a movement onwards from the more ancient, or rather that both have been arrived at by a movement onwards from the same system of contracting for marriage.

From the Book of Aicill (*Ancient Laws of Ireland*, Vol. III., p. 311) we learn that, among the Irish, the husband of a married woman had a right to her adulterine son till he was purchased from him by the actual father. The actual father, a man who could prove his paternity, had in virtue of his paternity, a right to purchase. But, until he purchased, and if he did not purchase, the child was the husband's. The father, having proved his paternity by the methods in vogue, and having made full payment—body-price, honour-price, the price of fosterage, and what had been paid on account of delicts—got the child for his own, and thereafter it was on a par with his legitimate children. Here there was the concession, in favour of paternity, of a right to purchase—that was the amount of difference between the Irish custom and the Hindoo. Paternity had gained the right to overcome the ancient effect of contract by means of contract.

From the same book (*Ibid.* p. 541), under the head of "Abduction without Leave," we learn that the child of a woman who had been abducted without leave had from her family, unless begotten more than a month after the abduction, belonged not to the abductor but to the mother's family. And if the mother had been forcibly abducted it belonged to them absolutely—they might refuse to sell it to the abductor. It was a moot point whether, if they wished to sell, he could refuse to buy. And if, as the result of such a state of circumstances, he in the end got the child gratis, the law obliged him to educate it, "because it is a good contract for him" (*Ibid.* p. 543). If the mother had consented to the abduction, the abductor could force her family to sell. In either case the abductor had to pay for his own child "the full price of its life" (*Ibid.* p. 541). It was, after purchase, considered as "the child of a first wife of contract, or of an adaltrach woman of contract." When there had been an abduction without leave, the woman's family were allowed a month "to force the man to the law" (*Ibid.* p. 403)—to bring him to terms about her or to reclaim her. When she remained with him longer, if there had not been a contract there ought to have been, and her family, for their neglect, lost the right to the offspring (*Ibid.* pp. 543, 403).

Here we see that contract about the mother was necessary to take her children from her family and give them to their father; that children begotten while there was no contract (if the mother's family were not considered to be in fault) belonged to the mother's family and had to be bought from them by the father—to whom, if he were altogether without excuse, they could refuse to sell; but that custom had fixed a time within which a contract ought to be come to, or the woman brought back. That the woman's family lost their right if they did not come to terms about her, or get her back, within a customary period, is the amount of difference between the Irish custom and the Hindoo. Abduction was common among the Irish, and people had settled that it was well to make the best of it.

That the right of the woman's family to her offspring was originally unlimited where there had been no contract about her—that it was older than the limitations on it—is not to be doubted. These limitations are only intelligible as having been made in the interest of social peace—though, no doubt, there was also a concession to paternity. But when the abductor was wholly in the wrong, and so long as the woman's family were considered still to be in the right, notwithstanding the man's paternity, the child belonged absolutely to the woman's family.

Among the predecessors of the Irish, therefore, as among the early Hindoos, the children of a woman who had not been parted with by her family always belonged to her family; and, on the same reasoning, the adulterine child of a woman who had been made over to a husband always belonged to her husband.

Sir Henry Maine only tried to show that, consistently with the Patriarchal Theory, the "fictitious" sons might have been sons "in some remote degree." His reasoning has not gone far enough to explain to us how they could have been sons in any degree; and it has been shown that the protective power theory on which it rests is simply excluded by Hindoo facts. In fact they were recognised as sons from the earliest times known to us; and that they were sons is accounted for by the principle laid down by Manu that a child belonged to the man who was "lord" of the mother.

That principle too has been shown to have underlain all Hindoo family custom. It gave the child of an unmarried woman to her father. It gave the illegitimate child of a married woman to her husband. It gave the legitimate child of a married woman to her husband. And it took his legitimate child by his wife from him, and gave it to his wife's father, in a case in which full right was not conveyed. Right to the woman then was the foundation of paternal right. And it is the principle that a child belonged to the "lord" of the mother, or rather the whole series of practices in which it was embodied, which had to be derived from the Patriarchal Theory.

It is however to a view of the growth of society widely different from the Patriarchal Theory that the Hindoo practices lend countenance—to reconcile them with that seems utterly impossible. Until the contrary of this is shown, it would not be open to Sir Henry Maine, even if other obstacles to his doing so were wanting, to say that the heirship of the son of an appointed daughter marks an initiatory step in the breakdown of Agnation. No case, indeed, presents more formidable difficulties to an upholder of the Patriarchal Theory than this one. For, besides that, when the daughter married, the husband, in virtue of the reservation made, lost the paternity of his legitimate child, when she did not marry (and there was no need that she should marry), her child was her father's because she was his. And it has been shown that this proves that the father's stock could be transmitted through her—though a woman. So serious a divergence from the Patriarchal view could scarcely denote an early stage in the breakdown of the Patriarchal Family. There was no difference in the world between a son, or the son of a son, and the son of an appointed daughter. The daughter's son was as good an heir as either, and delivered his grandfather in the next world as well as either. In Manu, too, the continuation of the family through daughters is represented as of immemorial antiquity among the Hindoos. "In this manner," it is stated (Manu ix., 128), "Dacsha himself, Lord of created beings, appointed all his *fifty* daughters to raise up sons to him for the sake of multiplying his race."

Going back now to the substituted begetting on widow or wife, it must be said that Sir Henry Maine appears seriously to underrate the facts. He tells us, the Levirate was permissible in the opinion of "some of the Hindoo doctors." And, "here and there," there was an extension of the same thing in what to modern thinking appears a grosser form. It is Hindoo practice that concerns us rather than opinions about it. But that Apastamba, as required by his dogma that the "son belongs to the begetter," disapproved of the Niyoga, seems to be all there is to say for the limitation in either of those statements.

Every other early writer includes the son "begotten

on the wife" in his list of persons recognised as sons, and most of those writers give him one of the highest places in the list. Generally he comes after the legitimate son and either after or before the son of an appointed daughter (to whom Manu, holding him as equivalent to a legitimate son does not give a separate place in his list); and his rights as an heir are carefully defined. Not one of those writers suggests a doubt as to his right to inherit, and when there was no legitimate son he was held fully equivalent to a legitimate son. Gautama, though he had a strong leaning to Apastamba's doctrine, compromised the principle that a son belonged to the begetter (or rather he gave it up) because he felt it necessary to leave beyond question the position of a Niyoga son. This shows that, instead of being a doubtful practice of exceptional occurrence, the Niyoga was thoroughly established. What better evidence of its recognition could one desire?

The recognition of the Levirate child was no less general than that of the son "begotten on the wife." And he had in every respect the position of a legitimate son—he was indeed born to be the heir of the dead man who was counted his father.

It is true that Manu regarded the practice of substituted begetting with disfavour, and talks of it as "only fit for cattle." But he prescribes the rule for carrying it out—enjoining at the same time that only one child or two at most should be begotten on the widow or childless wife. And, while contemning the practice, he discloses that even the "twice-born classes" had had it formerly. He speaks of it as of an ancient practice, which had died out—or was dying out—among the better sort of people. But he fully recognised it—no doubt because it was still too general not to be recognised.

But even what Apastamba says of the Niyoga bears out what has been said of it above. What he says is as follows:

"(A husband) shall not make over his (wife), who occupies the position of a 'gentilis' to others (than to his 'gentiles') in order to cause children to be begot for himself. For they declare that a bride is given to the family (of her husband and not to the husband alone). That is (at present) forbidden on account of the weakness of (men's) senses. The hand (of a gentilis is considered in law to be) that of a stranger, as well as (that of any other person except the husband). If the (marriage vow) is transgressed, both (husband and wife) certainly go to hell. The reward (in the next world) resulting from observing the restrictions of the law is preferable to offspring obtained in this manner (by means of Niyoga)."*

Were there nothing else known about the Niyoga this passage would be enough to show that it was an established custom in the time of Apastamba; that the Niyoga child was then a good heir; that it was then an old-established custom; and that the customary explanation of it was that a wife was given to the family of her

* Apastamba II., 10, 27, 2–7. *Sacred Books of the East*, Vol. II., p. 164.

husband and not to the husband alone. It also shows (which is less important) that, as his dogma required him to do, Apastamba disliked the Niyoga very much. He suggests that those who concurred in arranging for it would "certainly go to hell." But the last sentence of the passage admits that offspring, that is heirs, could be got by it in this world. Rewards in the next world are promised to those who resolve to forego such offspring. And it will be seen that, in the first sentence, the commentary overrides the text, changing the meaning and stultifying all that follows—obviously because what Apastamba had enjoined was so contradictory of all that was known about usage that nobody could take it as a statement of actual law. Apastamba is made by the commentator not to deny (or rather to admit) that the explanation made for the Niyoga had in earlier ages been a sufficient reason for it. And he sets it aside, not as being historically groundless, but as embodying a view of marriage unsuited to the weakness of his contemporaries, or, let us say, to human weakness.

That it was on this view of marriage the Niyoga was justified is the most valuable thing we get from Apastamba. He shows us that a practice which, so far as it went, was equivalent to Tibetan polyandry proceeded upon a polyandrous theory.

Sir Henry Maine ought, it would seem, to have reckoned with that theory. But after all, it is the account of the Levirate which he has himself propounded that concerns us here. And that has failed completely. He has, to speak plainly, shirked the difficulties of the matter. He has tried to show how it might have been held that "the family" was properly continued by the Niyoga child. What he ought to have shown was how, among a people who had had a strict view about paternity, it could be held that the line of a particular member of the family was properly continued by the Niyoga child. It being plain that that could not be held except where there was some degree of indifference about paternity, the question, what degree of laxity about it—whether much or little—would make the Niyoga child look "very like a real son," presented itself. But he has not considered it. The consequence is that he has reasoned the matter much as a Conservative Hindoo might have argued it against Apastamba. What he says would be of some weight from a polyandrous point of view. It has no weight at all from a Patriarchal point of view.

It may be well to add as to the Levirate among the Hebrews—since an explanation offered for the Hindoo institution has that to reckon with—that there can be no question raised as to its having been an established and an ancient custom. The story of Judah and Tamar shows that the childless widow of an elder brother could claim to have his brothers in succession for her husband; and that her child by her new husband took the place in the family of the husband who was gone, and was counted to him as a son. And the Levirate was retained in the law of Israel—the duty of a husband's brother to the childless widow was prescribed by the law—at a much later time when marriages between persons related by blood or affinity were in other cases strictly prohibited. The law took his legitimate child from the new husband and assigned it to the brother deceased, that it might be his heir. It was not every child of his that would suffice for the purpose. By itself the blood of the family was useless for this purpose. Among both Hebrews and Hindoos the child who could be assigned to the deceased brother was also a child of the woman who, had that brother lived, would have been the mother of his children.

It may by this time be permissible to consider it quite hopeless to connect the body of family custom found among the Hindoos with the Patriarchal Family. The interval between the two seems too great to be bridged by any effort of conjecture. For the Patriarchal Theory, as a hypothesis to account for the Hindoo facts, there is, it would seem, nothing to be said. And we are now in a position to feel the full force of the reasons against allowing Sir Henry Maine's assumption that the Hindoos, at an earlier time than any we know about, had the Patriarchal Family—as well as to see the weakness of the reason for making it. Let us state briefly how the matter stands.

The facts of the Hindoo family tell very strongly against the Patriarchal view. It would take very powerful reasons to bear down their evidence against that view, so far as the Hindoos are concerned. And to bear them down there is only a conjecture—propounded in connection with a subject of the greatest difficulty, the history of which has still to be properly investigated, and is at present much more obscure than the history of the family on which it is sought to make it shed light; but got in reality by means of the averment—which there are no means of supporting—that we constantly meet with the Paterfamilias "on the threshold of jurisprudence." To bear them down, that is, there is a conjecture for which there is really no basis, and which, though made in course of a disquisition on ancestor-worship, might as legitimately have been made without any reference to ancestor-worship.

Observe, too, the nature of this conjecture, as related to ancestor-worship. It is that the Patriarchal Family was necessary to give birth to ancestor-worship, and yet could not live with it. The Patriarchal Family was necessary for ancestor-worship to begin with, but too much for it to go on with. It would take some reasoning to show that it is not inconsistent with itself. What is perhaps more important is that it is so framed as to elude investigation.

Since it is made in connection with ancestor-worship it must be said that, before founding upon that, one would have to inquire whether ancestor-worship had not its rude beginnings—and, if so, what the family has been where these have been found; what the family has been also where it is found in full development; and whether it, in fact, appears to be specially connected with any one set of family relations and powers.

Without this having been done, there is no excuse for sweeping inferences from ancestor-worship.

In inferring the Patriarchal Family from (or in connection with) ancestor-worship, Sir Henry Maine has, of course, assumed that the family underwent no change all the time that ancestor-worship was growing up. He seems to imply that, after that, it deteriorated rapidly—the reason being that the Patriarchal Family did not yield children enough for the requirements of ancestor-worship.

His conjecture having been made specially for the Hindoos, let us see what difficulties there are in its way in the case of the Hindoos.

First, then, he asks us to believe that paternity was fully recognised when ancestor-worship arose. This is his first step towards the assumption of the Patriarchal Theory; and is meant to give him the Father, with a position undefined, but founded upon his paternity. He means by paternity physical paternity; and he means that it was recognised as the basis, or, at any rate, as a basis, of paternal right. Less than this would avail him nothing.

If we are to conjecture, however, is it not the reasonable thing to conjecture that the paternity which, among the Hindoos, was enough for ancestor-worship to go on with, was enough to set it going? The Hindoo father, such as he was—head of the household, protector, possessor of a father's customary rights, often, no doubt commonly, father in fact—would have made, it would seem, as good a basis for ghost-worship (if that is to be taken to have been the foundation of ancestor-worship) as any father could do. It is true that the legal basis (so to speak) of his parental right was not actual paternity, but the fact of his being "lord" of the mother. But it is not easy to see how, as regards that matter, this could make any difference. Why, at any rate, should we think it would have made a difficulty for ancestor-worship at the beginning, when we know that it made no difficulty afterwards?

But there is very strong reason against carrying conjecture further. We know that the high doctrine about paternity maintained by Apastamba was a novelty among early Hindoos, and that it found a practice hostile to it so firmly established that for ages it made no sensible impression upon custom — ancestor-worship going on all the time. We learn, however, that it began at once to influence conduct; and probably it influenced conduct greatly, even among those who did not accept it as a doctrine.

And what appeared to the supporters of this doctrine to give it a means of acting upon conduct? We have seen that it was ancestor-worship — that they should represent that the sonship of custom was insufficient for the uses of ancestor-worship. The new doctrine sought support from ancestor-worship and received it—though not enough for the overthrow of ancient customs which had long co-existed with ancestor-worship without any thought of their being unfit for its requirements.

Can we believe that a system which, as thus appears,

tended, when it was mature, to support paternity, undermined paternity when it was of newer growth—making men expect the benefits of sonship from persons whom (if physical paternity were thought important) they well knew to have no claim to be counted their sons? Then, can we also believe that, having done this, it replaced the system of sonship connected with paternity, not by a bundle of unconnected customs, but (this is the fact to be reckoned with) by a system of sonship consistently based upon maternity? This, too, while the high doctrine about paternity just spoken of appears to us among the Hindoos as a new doctrine, which protested against the ancient custom which gave children to the husband of their mother.

Are we to think, too, that, after ancestors began to be worshipped, descendants became scarcer? Did women become more immoral? Did men become more careless of their wives? There appears to be no reason for answering such questions in the affirmative; but, unless reason can be shown for answering them affirmatively, they all tend, and strongly too, to make us believe that, when ancestor-worship arose, the Hindoo family was no better than it was when we first come upon it.

More than enough has surely been now said to show that it is only while the facts are neglected that Sir Henry Maine's suggestion can be listened to.

What he seeks to get by it, as has been said already, is the Father, with a position arising out of his paternity. But, even were this conceded to him, it would leave him a long way short of that to which he moves on—the

Patriarchal Family, with Patria Potestas in the Father, and Agnation as the family bond. To this immense enlargement of his assumption ancestor-worship does not contribute anything; nay—what by itself would exclude the Patriarchal Theory—the Hindoo observances spoken of as ancestor-worship were paid not only to male ancestors but to female.*

* There remains a question as to the relation between ancestor-worship and the observances paid by Hindoos to deceased ancestors. These observances were, according to the books, intended to deliver the deceased from the hell called Put.

CHAPTER XVII.

SONSHIP AMONG THE HINDOOS.

Having been led so far into discussion of the right to children among the Hindoos, it seems worth while to go further, and inquire to what early system of counting kindred the Hindoo customs are really related. Let us first recall the facts as they have appeared to us.

Among the early Hindoos the child did not belong to the begetter. A man had no right to a child merely as being its father. There was affiliation of children to the mother; and children, following the mother, were also affiliated to the man who was "lord" of the mother. Sonship was not founded upon paternity; and the family did not include the illegitimate children of the husband (as we have seen that it might do among the Irish), although it did include the illegitimate children of the wife. It was founded upon maternity; and, through the mother, a child belonged to the man who possessed rights over the mother. The "lord" of an unmarried woman was her father; and she was affiliated to her father and of his stock and family. Her issue also were affiliated to her father, and of his stock and family. Marriage passed a woman, and with her usually, it would seem, her issue, if she had any (no doubt by some process like dedication), into the stock and family of her husband. It transferred her father's rights over her to her husband, except that, in one case, something was usually reserved. Her husband was now her "lord," and, saving the reservation, her issue was his—of his stock and family and affiliated to him. The reservation of a son of the marriage, customarily made in the case of a brotherless daughter, kept in the family of the woman's birth a child which, but for the transference of rights made at marriage, would have belonged to it. It suggests that to part with the woman was not *necessarily* to part with her issue.

In general, however, as has just been said, the husband not only got the woman's issue born after marriage but, it would seem, her issue born before. Her child born before marriage—before her husband's "lordship" had begun—passed with her out of her father's stock and family into her husband's, and was thereafter affiliated to her husband. And, on the other hand, her issue born after the husband's death—after actual "lordship" had ceased—could be counted his with the help of the Levirate. Custom, in times comparatively late, retained a mode of replacing him for one case—that of his dying childless; and, substituting for him his brother, assigned to him, though dead, as his heir the child of the woman who had been made over to him in marriage.*

It may be added that, incidentally, reason has been shown for concluding that marriage was commonly among early Hindoos (as among other peoples in a rude state) an affair of sale and purchase; and that the peculiarities of Hindoo family law sprung out of an early system of contracts for marriage made on that footing. The facts indicate that the ordinary contract transferred to the husband the whole right of the woman's family—the right to her and her issue, including any child there was at the time of transfer; but that her father usually reserved the right to one child when he had no son—and perhaps the right to claim one in the event of his afterwards finding himself sonless.

It is easy to see that the facts are exactly what we might expect to find were such a system of contract superinduced upon the system of counting kinship through females only.†

* A woman who, were the man alive, would be the mother of his children was indispensable for the Levirate. If he left no widow, a man could have no Levirate heir. Custom afforded a means whereby, when dead, he could be replaced, and so that the child could be assigned to him as if he were living. With Tibetan Polyandry, brothers are interchangeable with, or equivalent to, one another; the child of one is the child of every other—while every child is reckoned in a special sense the child of the eldest brother. The Niyoga child is defined by Manu as "the lawful child of the wife."

† Of sons engrafted upon the family to supply the lack of children, of whom there were five varieties—the son given; the son made (adopted); the son rejected by his parents; the son bought;

This system of kinship makes children of their mother's stock, and affiliates them to their mother's family—whatever be the connection of which they are the offspring. What would happen then if there were superinduced upon it a system of contract in virtue of which a woman with her issue was transferred from her family to a husband, and she was taken into the husband's stock? The children, being of the stock of their mother, would now be of the stock of her husband—so that they could be his children; and, being his in virtue of the contract, they would be his children. And as his daughter would be his, so would her issue—except so far as he had transferred it by contract to another man. Her child born before marriage would be his. If, when she married, he bargained to have a child of the mar-

and the son self-given—no account need be taken here. For the natural family in some form must have been established before there could have been thought of admitting such persons to the family. It is worth while to note here, however—with reference to the position of the mother in the Hindoo family—a fact not mentioned in its proper place, that, in the statements made about these classes of sons in Manu, it is involved that the mother as well as the father had a right to and a share in disposing of children. The son given, according to Manu, might be given by the father or by the mother. Kulluka qualifies the statement as to the mother by the words, "with her husband's assent." But, even with this qualification, the mother had such a right over her children that she could take the initiative in giving one away. And, of course, when there was no husband her right was unqualified. Similarly, it appears that the son taken as having been rejected by his parents must have been left without either parent to take care of him; while the son bought is defined as a child purchased "from his father and mother," and the son self-given as "he who has lost his parents or been abandoned by them." (Manu ix. 168–177.)

riage, that would come back to him. When he gave her over to a husband without reservation, her children would be the husband's. And all her children would be his—if he chose to have them—whether they were his own offspring or not.

Contract, following upon kinship through females only, would therefore yield in the first instance precisely the system of parental right and sonship which we have found among the Hindoos. The circumstances seeming favourable to the actual paternity of the husband, a strong preference for legitimate children might have been expected quickly to spring up. And it is, perhaps, surprising to find the wife's illegitimate and adulterine children in plain terms assigned to the husband after Hindoo customs had been long recorded and had become subject to rigid criticism. There is no doubt of the fact, however, and we know the theory which was held to justify it. Combining both with what has just been said, it seems pretty clear that it is not Agnation but the system of counting kinship through females only that is to be descried as operative among the prehistoric Hindoos.*

* It is proper to state that, though Mr. J. D. Mayne believes that the Hindoos had the Patriarchal Family, his account of sonship among the Hindoos, which has already been referred to, is up to a certain point substantially coincident with that given in the text. For the case of the Levirate son, he has given up his theory of sonship. But apart from that, such difficulty as he has found in accounting for the facts has arisen out of his overlooking that marriage and its consequences must have been founded upon contract, and that the consequences would have been unusual with a condition (express or implied) which became operative in exceptional cases only. The

To mention Aryans and Africans together seems shocking to some people; but the probability of the explanation of Hindoo facts which has just been offered cannot but be increased by our finding that it appears to be contract superinduced upon female kinship which produces in Africa results very like those which so much astonish us among the Hindoos. An explanation which involves that kinship was once counted through females only can be more readily accepted for African facts by themselves than for Hindoo facts by themselves—since in Africa it is a common thing to find kinship taken only through the mother. And, given a set of Hindoo facts and a set of African facts which are much alike, it is reasonable to feel our confidence strengthened in an explanation offered for the former, if we find

case of the son of an appointed daughter is at least as easy of explanation as any other; but he has failed to explain it satisfactorily, because his thoughts were fixed upon the husband's right instead of on the contract out of which it sprang. What is more important is that, through treating the husband's right as if it were an ultimate fact, he has been kept from carrying analysis of the facts he was dealing with as far as he might have done, and led into entirely misapprehending their significance.

It should be noticed, too, that he has—influenced apparently by a reminiscence of the "Family founded on Power"—used the term ownership to denote a man's rights over his family. A more general term is preferable to one which can seldom be literally applicable except in a transitional state of family relations. But more than that, there was absolutely nothing in the Hindoo facts to press the suggestion of ownership upon him, and he did not need the term for any purpose of explanation. It cannot be necessary to illustrate the fact that a wife who has been got by purchase and her children are usually not slaves.

that it is probable enough as an explanation of the latter.

The African cases in which children "follow the mother"—belonging to her clan, bound to her religious observances, heirs in her family, and not heirs of their own father—are, the principal of them at least, well known, and are too numerous to be here recited. Nor is there any need to repeat what has been said in a former chapter* to make it appear that there has been a change in kinship—though there can be no harm in giving a reminder that kinship can be found in stages of transition, and that we have seen in use expedients well fitted to effect a change in it, one of which, with whatever objects, was in use among the most famous of ancient peoples. We may confine ourselves to certain African facts which appear fitted to throw light upon the influence which contract has had upon relationship. Similar facts, it may be said, could be brought together from many other quarters, and the argument to be submitted greatly strengthened by them.

When, in Africa, a man on marrying enters his wife's family (beenah marriage), he usually gains his footing, as Jacob did his, by service. In that case, the children all belong to the wife's family; and if the husband goes away—which he is free to do at any time—he can take neither wife nor children along with him. His being father avails him nothing. His having been husband avails him nothing. The children "follow

* Chapter xiv., "The Origin of Agnation."

the mother" and belong, as she does, to her family. Very often however—indeed, it may be said commonly—the communities which still keep up this kind of marriage do not practise it exclusively. Families are more or less willing—for value received—to give a man a wife to take, or let him take his beenah wife, to his own people. The contract, in this case, usually transfers to the husband the woman and her offspring; it is this which is bargained for; and where this has not been bought and paid for, even when the woman has been allowed to live with her husband in his own village, we find that the children may be claimed by her family. If payment has not been made, the woman's family have not—to use the current phrase—"given her up." Here a contract which carries away children, with their mother, from the mother's family is seen to be the sole basis of the father's right to his children.

This need not mean that men are indifferent about the paternity of their children. They may be either careless or very particular about that. And the circumstances show that men have a strong desire to have their offspring for their own. It means that paternal right is not, as with us, a consequence of a man having children born to him in the *married state;* that, moreover, it does not arise out of paternity; and that it arises out of purchase. And, arising out of purchase, it extends to all issue of the wife. Paternal feeling it is, no doubt—and the wish to have heirs in his own household—that makes a man desire that his children should belong to himself and not to the family of their mother. But it is through contract alone they come to belong to him. They "follow the mother," and—contract apart—are of the mother's family. Contract takes them, with their mother, out of the mother's relationship, and makes them over to their mother's husband. It is the means whereby children are separated from the mother's family, and "given up" with her to her husband.

We often are not told, and cannot be sure, whether the African peoples among which we find beenah marriage are exogamous totem kindreds or not—though there seldom is serious doubt about the matter; nor are we often told whether different kindreds are interfused in the same village, or whether the village contains only one kindred. Kindreds of some sort, however, there are among them. And we have seen how they are recruited. A kindred or family is continued through women by means of children born to women about whom no contract has been made. It is continued also through men by means of contracts by which women and their issue are "given up" to them. It can as a matter of course, however, be continued through the women. It is by means of an expedient it can be continued through the men. For men, blood-connection and marriage, by themselves, go absolutely for nothing. There must also be purchase. On the other hand, there is by birth a connection between children and their mother which, if undisturbed, makes them of the kindred to which their mother belongs. The connection between mother and child is acknowledged and never ignored. The connection between father and child is frequently ignored, and the acknowledgment of it has always to be paid for.

Here it is obvious that blood or stock connection does not count for so much as in the cases—which are more numerous in Africa—in which, notwithstanding that the wife is purchased, her children are inalienably of her kindred, and of her kindred alone. But, again it is equally clear that when children are alienable from the mother's kindred to her husband the desire of men to have their children to themselves counts for a great deal. And is not this enough to account for all the difference there is between this class of cases and the other class? The extent of the difference is that instead of children being necessarily of the kindred of their mother, they are taken out of her kindred if the right to children has been purchased by her husband.

Plainly, where this right can be got by purchase, and arises to men in no other way, the extent to which children do not belong to the kindred of the mother, as it depends entirely upon the willingness and the ability of men to purchase the right, is governed by that which makes men willing to purchase it, and has established a custom of purchasing it—and cannot be greater than what that will account for, though (since the ability may be wanting) it may be less. Now that which disposes men to purchase the right is the desire to have their children to themselves. And, in fact, among the peoples we are speaking of that desire is more common than the ability to gratify it.

The extent, therefore, to which, where the right to children can be purchased by the husband, children do not belong to the kindred of the mother, is amply accounted for by the desire of men to have their children to themselves.

Observe now that, if that desire were to cease, purchase would cease, and all children would belong to the kindred of their mothers. And, again, if there was a time when the desire was without the means by which alone it becomes operative, children then belonged inalienably to the kindred of their mothers.

It cannot be believed that contract has been an original, or that it is anywhere more than a temporary, basis of kindred-connection. And, therefore, we may conclude that where now fathers can get the right to their children by purchase only, children did originally belong inalienably to the kindred of their mothers.

Moreover, it is not to be doubted that the connection so clearly acknowledged between mother and child implies a recognition of the blood-connection between them. And, through it, children still accrue to the kindred of their mother—unless contract comes in to carry them off to her husband. When they always accrued to the family of their mother, therefore, kindred connection was based upon blood-relationship through the mother. We may advance, indeed, to this conclusion without a word provided we are not prepared to hold that blood-relationship has never been thought of at all among the peoples which practise beenah marriage.

It scarcely need be pointed out that the mother's blood, supposing it was anciently the only basis of kindred connection, could not but dwindle in importance in relation to it in the proportion in which contract grew into importance in relation to it. It was inevitable that, as the one came into prominence, the other should lose prominence; and it is not surprising that, when the one encircles and envelopes the other, as it does now—ideas derived from contract prevailing—the relation of a woman's family, as well as that of her husband, to her, and to her children when young, should sometimes resemble mere ownership.

It appears then that we are justified in believing that blood-connection through the mother was, until contract was brought in to cut it down, the sole basis of stock-connection in the cases we have been considering; that, in the affiliations of children which we find in them, we have a sequel to the system, so common in Africa, of counting children of their mother's blood and clan; and that, in the system of contract in use, we see a means by which the limitations of the old kinship were got rid of and the way was prepared for a new one.

But even where marriage is not beenah, and never takes place without purchase, where the wife goes to live with her husband, and the children need not belong to the mother's clan, we find in Africa that marriage does not in all circumstances give the children to the husband. Among the Makololo, the price paid on marriage might merely cover the right to have the wife; and in this case the children belonged to the family of the wife's father. It might cover a certain right to the children also, if that had been contracted for; but never such a right as completely separated them from the mother's family. To effect this, it was necessary that a further price should be paid at the mother's death. This paid, her family had "given her up," and her children were entirely dissevered from them.

Here, with a general desire on the part of men to have their children for their own, we see contract overcoming, as it would seem, with difficulty—not at a single operation—the connection of children with the relatives of their mother. And it appears, if possible, more plainly than in the former class of cases that the husband's paternity could not make the children his children; and that mere marriage left children of the family of the mother. Again we may conclude—even more confidently than before—that the affiliation of children, in case that the dissevering process employed had not been carried out, to the family of the mother was a remainder of the system which counted them as necessarily of the mother's blood and clan; and that we see the father moving towards the full position of an acknowledged parent by means of a system of buying off the mother's family.

It may here be said that among many African peoples which count kinship only through the mother, the father, who gets his wife by purchase, has a sort of property right in the children—while, that is, they are young and helpless. The wife's family also have rights in them; and their protection, the fear of their resentment—that is of the blood feud—effectually distinguishes children from slaves. We sometimes find, however, that the husband can pawn the children. We learn in other cases that he may not (or perhaps it may be that he dare not) sell them unless he can get the consent of his wife's relations; while, if he can arrange with the latter, there is nothing to hinder him from treating them as absolutely his property. From this it may be understood how, when a desire to have the children of their marriages to themselves had become general among men, effect should in many cases be given to it by means of contracts with the wife's family. Where religious observance had become prominent in connection with clanship, the dedication of the wife—of which, as we have seen, a beginning was made in Guinea in the system of bossum-wives—might also be necessary. When contract alone sufficed, as the first effect of it was to free the father from all claim on the part of his wife's relations to her children, there is no difficulty in understanding how, for a time, ownership should, in appearance, be more prominent in connection with relationship than blood-connection.

That contract carried to the African husband, as far as could be done, the right of his wife's family—the right to all the offspring of his wife—might be shown by many examples, but will be sufficiently illustrated by two, taken from peoples with very different arrangements as to kinship.

The Ashantees, as we learn from Bowdich, counted kinship only through the mother. When, among them, a man had been missing for three years, his wife might be married again, and, if thereafter he returned, the second husband's claim to her was good. But the children of the second marriage were the first husband's as much as if they had been his own children, and accordingly, they might be pawned by him. The right over the woman's children which marriage had given to the first husband extended, therefore, to all the children she might have, and it survived the second marriage of the woman; so that, though he could not recover her, the sort of property interest he had in her children was unimpaired.

Equally conclusive as to the right carried to the husband by purchase, and more interesting in relation to the present inquiry, is the passage subjoined, which refers to a branch of the Bechuana people of Southern Africa:

"Macheng was not born till some years after the death of Khari, his reputed and 'legal' father. Neither does this affect his title to be chief. Khari, having elevated the mother of Macheng to the dignity of head wife [it is explained in another place that she was the wife of highest rank, and that her children, therefore, had a preferable claim to the succession], and having paid her price in cattle, she and her offspring are to be reckoned to Khari although the children should be born a dozen years after his death. It is not etiquette ever to refer to the man who thus 'raises up seed' to another in connection with such children. They are the children of him who is dead. But when Sekhome

[Khari's eldest son, who had usurped the chiefship] was vexed, he sometimes sarcastically declared that his rival was the 'child of cattle,' meaning that the price paid for Macheng's mother at her marriage with Khari was her son's only title to the chieftainship. There is never any question among the natives, however, as to the validity of this title. Even the most ardent friends of Sekhome admit that, according to their customs, Macheng is the rightful heir." *

This passage reads as if it had been written to show the identity of the family law of the Bechuanas and the family law of the early Hindoos—of which probably the writer was entirely ignorant—the principle of each being that a man was "father" of all the children of his wife. After reading it, there cannot remain a doubt that the peculiarities of the Hindoo law sprang out of a system of contract for marriage—that is, a system of purchasing the wife and her issue. And the marriage law of the Bechuanas was in principle identical with the marriage systems already noticed, as to which reason has been shown for concluding that they show a progress from the system of counting kinship only through the mother.†

We see no appearance of ownership as a title to children among the Bechuanas, however. There is affiliation to the father. And there is even more

* *Ten Years North of the Orange River*. By John Mackenzie. Edinburgh. 1871. Page 364.

† If space permitted, some striking evidence to corroborate this conclusion could be produced from among the Bechuanas.

distinctly affiliation to the mother. The connection with the mother carried so much weight that the children of a polygamous husband took rank among themselves from the mother—their maternity thus seeming to overshadow their paternity. It comes out very plainly too, that the husband's connection with his children arose out of contract—out of his having become by purchase the husband of their mother; and that what the contract carried to him was the right of his wife's family—the right to all the woman's offspring. For, in virtue of the contract, his wife's offspring could be counted to him as children even after his death; and counted to a chief so that the child of his principal wife born "some years" after his death was, by universal admission, his lawful heir. This, too, was not a consequence of the deceased being childless, but held good while there were surviving sons of his by inferior wives born in his lifetime. But here some explanation is necessary. In the passage which has been quoted it is hinted that there was a particular man who "raised up seed" to the deceased; and the writer mentions in another place,* as one of a number of Levitical practices found among the Bechuanas, "the custom of raising up seed to a deceased brother or relative;" and it may therefore be taken as beyond doubt that it was as the issue, or reputed issue, of a Levirate connection that Macheng was heir, or that other children of the widow would be counted children of her dead husband.

It was none the less in virtue of the contract which

* *Ten Years North of the Orange River*, p. 393.

transferred to Khari the right of his wife's family to his wife and her offspring that her child born after his death could be counted his. Macheng, as his rival said, was "the child of cattle." In virtue of the contract, the woman's children were to be counted to the husband as long as it was possible so to do. Some means of making the right conferred by marriage available after the husband's death was, no doubt, indispensable. And such a means there was in the Levirate—in the custom in virtue of which the deceased could be effectually replaced by his brother or other near kinsman.

It should not escape remark that, finding the Levirate among the Bechuanas in an early form—not cut down by the endeavours of generations of priests and lawyers to modify it in the interest of decorum—we in this case find it employed not to give a childless man an heir, but to enable a man's head wife to give him a child who should cut out from succession his sons by wives of lower rank. And, purchased woman as she was, it is a right of the wife alone which is disclosed to us in this particular case. It does not appear whether there were any customary restraints upon the Levirate among the Bechuanas, or whether there was any limit to the number of Levirate children.

As the Levirate is common to the Hindoos and to the Bechuanas, and as—if it must be deemed to be of polyandrous origin—we may, upon slight indications otherwise obtained, and indeed without any other indication, believe that, where it occurs there must originally have been kinship through females only—and may accept accordingly an explanation in which that is involved—some further observations may now be made upon the passage quoted above relative to the Bechuanas. It should be borne in mind that the Levirate was practised among the Hindoos (as we learn from Apastamba) upon a polyandrous theory; and that the attempts to account for it among the Hindoos independently of polyandry which we have had to notice have appeared to be entirely unsuccessful. It should be borne in mind, too, that both polyandry and kinship through females only have been very common in Africa; and that where, with kinship through females only, polygamy is practised to any extent, the two are necessarily found together. Besides this, it may be pointed out that, at any rate the motherhood of a child must have been regarded as more important than the paternity of the "father" wherever the Levirate has prevailed.

Apart, then, from our knowledge that the Bechuanas had the Levirate, finding that among them a child born to a woman years after her husband's death could be counted as beyond question the husband's lawful heir, to the exclusion of his sons by inferior wives, one could not have doubted that it was so counted in virtue of some recognised connection, and was not a mere child of chance.

The husband had, no doubt, as Mr. Mackenzie says, acquired the right to his wife's offspring. But, in ordinary course, the rights which a man possesses exclusively of other persons either cease at his death or pass then to some other person. Rights that are not exclusive, on the other hand, on the death of one of the sharers go on subsisting for the survivors. We know of a class of cases in which, with monandry, that is, an exclusive right in the husband, the husband's rights over his wife pass at his death to another person. They are found where the wife is regarded as property. In these cases she passes, with the estate of the deceased, to his heir. And, of course, we know of cases of marriage in which, the husband's right being exclusive, it ceases at his death altogether. This happens in all the more advanced cases of monandry; and it happens also in many of the ruder cases of monandry, notwithstanding that in these the husband gets, as the Bechuana husband did, rights over all his wife's offspring.

When, therefore, it had appeared that, among monandrous Bechuanas, the husband's exclusive right did not pass to his heir at his death, and yet was not extinguished, we should have been ready to conclude that, among the Bechuanas, there was some traditional means of giving effect to it, and, finding effect given to it so thoroughly, to conclude that this was not anything of the nature of a contrivance or expedient, but some connection which was not only sanctioned by custom, but which flowed naturally out of ancient practice. The alternative would have been to think that, in virtue of the right the husband had acquired, *any* child the wife might have after his death could be his heir—to suppose, that is, that the husband's right was gone, but that there remained in its place an unlimited right for his widow—which, as a supposition, must have been deemed extravagant.

Having concluded that the husband's right was made available by means of some recognised connection, one might have gone on, not unhopefully, to speculate as to the nature of that connection. For, in there being such a connection, it is involved that there was some man who, in virtue of ancient practice, could take the deceased husband's place, and who filled it to the common apprehension so sufficiently that the widow's children could continue to be deemed the children of the dead man. Since we know the Levirate as a means of raising up children to a dead man, and history discloses no other means for that purpose, it might with much probability have been concluded that the Bechuanas had the Levirate.

Might it not, at any rate, have been inferred—since the husband's right did not pass to his heir or cease at his death, but was carried on as if there had been some sharer in it—that the ancient practice indicated was one in which men had been sharers in marriage? Polyandry having been anything but rare, and having, in particular, been quite common in Africa, a polyandrous solution would have been highly probable. And what other would there have been worth mentioning?

As the Bechuanas actually had the Levirate, and in connection with it there is disclosed to us, more prominently than usual, a right of the wife—a right to have children by her husband's brother or kinsman, who should be counted her husband's children, and take the place in his family they would have had if born to her in his lifetime—how, let us ask, is such a right reconcilable with monandry? How, with monandry, could marriage give a right to a woman over her husband's brother—or, for that matter, a right to her husband's brother over her? And how could her children by the living brother be counted her children by the dead one? It seems clear that a monandrous solution is, in this case, not to be looked for. The more hopeless a monandrous solution of the Levirate the more confidently may we accept a polyandrous solution. But apart from that—taking the Bechuana case by itself—every part of Southern Africa being liable to strong suspicion of polyandry, it will be found almost impossible to rebut the suggestion it gives that the reason why one brother could so effectually replace another was that brothers had at one time been co-husbands, so that the death of one left the wife the others—that there had been a practice of polyandry in the Tibetan form, in which a man's child is also the child of his brother, while all the children of a brotherhood are counted specially the children of the oldest brother.*

At any rate, Sir Henry Maine's suggestion that the Levirate was a contrivance for giving a childless man an heir might be held to be excluded on the strength of

* The childless widow had the right to have her husband's brother for husband among the Jews. The widow had the right of initiative among the Hindoos also.

the Bechuana case alone. For all cases of the Levirate are so much alike that it is allowable to make each of them throw what light it can upon the origin and early conditions of the institution. And it was not necessary among the Bechuanas, for bringing the Levirate into play, that the husband should have died childless. In the Bechuana case, what called it into operation, or was the reason for it being put into operation was, in this particular instance, the childlessness of a particular wife.

Having found that among the Bechuanas, polyandry in the Tibetan form (or possibly in some closely allied form) seems to give the only possible explanation of the Levirate, and gives for that particular case an explanation of it, otherwise highly probable, it seems now worth while to recall that (to say nothing of the direct evidence of polyandry) the Levirate is not the only Hindoo fact which indicates a prevalence of polyandry in the Tibetan form among the early Hindoos.

There is a provision in Manu—not taken into account in either of the works which have been noticed in the preceding chapter—which ought not to escape the attention of any one who attempts to give, consistently with monandry, an explanation of the Levirate. Where any one of several brothers of the full blood had a son, "Manu pronounces them all fathers of a male child by means of that son"*—so that, adds the commentator, if such a nephew would be heir the uncle had no power to adopt a son. With Tibetan polyandry,

* Manu, 9, 182.

the brothers would all have been "fathers" of the child, and he would have been heir to all of them. And this rule regards them as his fathers in such good faith that it cuts the uncles off from getting by adoption sons who might exclude him from inheriting from them.

Have we here a remnant of ancient custom, that is, a relic of Tibetan polyandry, or a fiction, or how else is the rule to be explained?

Perhaps there is nothing Hindoos were more unlikely to do than to resort to a fiction or an arbitrary rule to prevent men from having each his own heir. To multiply heirs was, in the case provided for, to multiply households, and with households to multiply religious observances; and that was the interest and desire of every Hindoo "authority." And what reason could there be for the introduction of such a rule—if, indeed, there was any power capable of introducing it?

The rule we may take it then is beyond question old, and derived from ancient practice. It is intelligible that it should be found among a people governed by custom, among whom, in times past, it was common for a man's wife to be the wife of his brothers, and his child to be the child of his brothers; and that it should continue to be authoritative after it had ceased to be popular, or after the reason for it had lost substance. It embodies the polyandrous view of fatherhood and sonship. The Levirate embodies the polyandrous view both of sonship and of marriage. Combining the indications of the two, we seem to be very free from risk of error if we conclude that the early Hindoos were polyandrous.

If it must be concluded, as to both the Hindoos and the Bechuanas, that they were commonly polyandrous before they were commonly monandrous, and if an explanation offered for the family system of either, which appears sufficient as a hypothesis, involves that they at one time acknowledged kinship through females only, it should be very easy, as has been said already, to accept that explanation. For, wherever polyandry can be found, even in the Tibetan form, a prevalence—either with that polyandry or earlier—of the conditions which admit of kinship through females only is highly probable. But as to the Bechuanas, there should be no difficulty at all. For, besides their polyandry, their family system links on to African marriage customs which appear clearly to have been arrived at by a progress from the system of kinship through females only. And an explanation which is good for the Bechuanas, should be held good for the Hindoos.

We are carried back then to a time at which, among the Hindoos, the bride, obtained by purchase, was really given to the family—commonly enough, and during a period which lasted long enough, to enable the arrangement to leave marks upon common practice; and, beyond that, to a time when kinship being counted among them only through women, their family conditions were such as are found with that kinship—marriage which left children of the stock of their mother, beenah marriage, Nair polyandry. It may here be recalled that, in the marriage law of the Hindoos, as stated by Kulluka, we have already seen reason for believing* that children among them, previously to their being counted of the stock of the father, had belonged exclusively to the stock of the mother. And it should not be forgotten how secondary, relatively to maternity, was the part of paternity in creating family relations among the Hindoos.

How far, if the views submitted in this chapter be deemed maintainable, the inference of a prevalence of kinship only through females can be established by means of them for other peoples besides the Hindoos cannot be here considered. For the present it is enough to say that we may feel sure there is equal reason for that inference whenever it can be shown that the father's right to children has originated in purchase, and we find also the Levirate, or some other custom or institution of equivalent significance.†

* Chapter xiii., "The Origin of Agnation."

† It may be well, however, to mention that some of the most characteristic Hindoo facts of which the argument of these chapters has taken account, reappear in other nations where there is independent evidence of the importance attached to kinship through women. As to this, Professor Robertson Smith supplies the note subjoined:

"The ancient Arabs," he says, "recognised the right of the husband to choose another man to beget a child for him on his wife. (Bokhārī, 'Kitab an-Nikāh,' vol. vi. p. 127 of the Bulak vocalised edition), and in this case as among the Hindoos it appears that the object was to obtain a more excellent offspring. Compare Wilken, *Das Matriarchat bei den alten Arabern* (Leipsic, 1884, translated from the Dutch), p. 27, *et seq.* It is instructive to observe that this notice is given by Bokhārī along with examples of polyandria proper. Among the Germans, again, the husband who was not himself able to

The African cases we have noticed have been referred to simply as showing the influence of contract upon relationship. It need not be supposed that the history of the change of kinship among the Hindoos was precisely parallel. In Africa, the process of change has been very protracted, and the position of women is now in many cases very low. On the other hand, there are facts which tend to show that when the change began among the Hindoos the position of women was very high, and that their position continued to be high after the family system among them had become substantially identical with that which is disclosed to us in their law-books.

have issue by his wife was allowed to seek a substitute, a practice which Jacob Grimm, *Deutsche Rechtsalterthümer*, i. 443, *et seq.*, has shown to have long survived in the written customary law of the peasantry in various districts. Grimm calls attention to the fact that the same thing was allowed in Sparta and Athens, among populations where the evidence for primitive polyandry and female kinship is particularly strong and clear."

That the Germans once had the system of kinship through females only is indicated unambiguously by the evidence noted above, p. 253, *et seq.*

CHAPTER XVIII.

THE TUTELAGE OF WOMEN: THE HEIRSHIP OF SLAVES.

In *Ancient Law*,* Sir Henry Maine says he thinks it can be shown that the family, as held together by the Patria Potestas, is "the nidus out of which the entire law of persons has germinated." He then proceeds to explain the relation to the Patria Potestas of "the institution known to the oldest Roman law as the perpetual tutelage of women;" and of the capacity which, from the testimony both of ancient law and of many primeval histories, the slave appears to have possessed of being, under certain conditions, the heir, or universal successor, of his master.

This part of his subject is not treated thoroughly, or with full argument. While he specifies none of the cases in which a slave might be his master's heir, the few examples of systems of female guardianship which are referred to—he mentions the Hindoos, the Scandinavian nations, and "the invaders of the Western Empire"—appear to be those of peoples for

* Page 152.

which he believed himself to have already established his theory by means of Patria Potestas or Agnation. It may almost be doubted then whether this portion of his work was not inserted as being in itself curious and interesting, and consistent with, or corroborative of, what had been sufficiently made out before rather than to furnish fresh evidence of Patria Potestas. And it is quite clear that, if Agnation had been connected with Patria Potestas, as he supposed it to be, and had been, as he said it was, discoverable nearly everywhere, further evidence of the prevalence of Patria Potestas would have been superfluous—while, on the other hand, it is easy to see that there could be no hope of establishing the Patriarchal Theory by means of the tutelage of women and the heirship of slaves alone.

Nevertheless, the connection alleged to exist between Patria Potestas, on the one hand, and the tutelage of women and the heirship of slaves on the other, must here be discussed. But the discussion need not be a long one. For a great part of what has in this work been said of the connection supposed to exist between Patria Potestas and Agnation, applies also to the connection alleged between it and the tutelage of women and the heirship of slaves, and it will suffice to refer to that very briefly.

What Sir Henry Maine has said to show the connection of the tutelage of women with Patria Potestas is not unlike what he had previously said of the connection with it of Agnation. But there is a difference. Agnation, he said, was "as it were a mould," which retained the imprint of Patria Potestas. And of perpetual guardianship he says that it "is obviously neither more nor less than an artificial prolongation of the Patria Potestas when for other purposes it has been dissolved." Agnation, however, is described to us as a natural growth, as a necessary result of Patria Potestas; whereas we are told that perpetual guardianship was "a peculiar contrivance of archaic jurisprudence" for retaining women in the bondage of their families.* Agnation, moreover, is presented to us as a remainder left by Patria Potestas, having grown out of Patria Potestas and survived it; while of perpetual guardianship we are only told that it was a contrivance by which Patria Potestas, while it still existed, was artificially prolonged. Nothing is said to show that what suggested this contrivance, or made recourse to it natural, was Patria Potestas further than that this is obvious.

That this is obvious in the sense of appearances proving it seems, however, by no means to be the case.

There is no need to repeat what the powers of a Paterfamilias were. On the other hand, we do not know that at Rome guardianship at any time extended to a woman's person. So far as we know, it extended to her property only, and even to her property in a very limited degree; and it might be a sufficient account of it, as it existed at Rome, that it was intended to put a check upon alienation or waste in the interest of the Agnates who were the woman's heirs, and who,

* *Ancient Law*, p. 153.

in the absence of testamentary provision, were themselves the guardians. At any rate, comparison between the guardian's powers and those of a Paterfamilias shows that the former were slight while the latter were unlimited; and that the powers which were slight were confined to dealings with property, while the unlimited powers applied both to property and person. The one set of powers then was of an entirely different order from the other; it is not resemblance, but difference between them that is apparent; and, indeed, the two coincide only in this, that the persons who were subject to the restraints of guardianship were persons who had once—like all other persons—been subject to, and who had escaped from, the vastly greater powers of the Patria Potestas. That the powers of guardianship and the Patria Potestas both belong to the law of persons, and that, of this branch of the law, the latter was by far the most important title, appears, in fact, to be all that can be said for the suggestion that the tutelage of women, as it prevailed at Rome, was obviously a prolongation of the Patria Potestas.

Sir Henry Maine, indeed, assumes that the powers of guardians at Rome were originally more extensive than they are known to us to have been, and that they had been cut down while the Patria Potestas yet remained intact. Control of the woman's person, he tells us, was "apparently quite obsolete" at Rome. The assumption, be it observed, is not consistent with the guardianship of women having had greater powers of lasting than Patria Potestas.

If, however, we suppose the powers of a guardian to have originally approached nearer to those of a Paterfamilias at Rome, no doubt there would be more reason than there actually is for saying that it is obvious that the former were a prolongation of the latter. But what is there to justify the supposition? Nothing—except that guardianship can be found extended to a woman's person in places where there is no Patria Potestas, and the Paterfamilias is unknown.

Even in those cases in which they have been greatest, however, the powers of the guardian cannot be likened to Patria Potestas; and it cannot be said to be obvious in any case that the one set of powers has been fashioned after the other. On the contrary, if it had to be granted that, when a substitute for the paternal control came to be devised for women, the powers actually possessed by fathers at the time furnished the model for it, there would be a complete absence of reason for thinking that those powers were anywhere Patria Potestas. And perhaps in no case is there less appearance of connection between the two sets of powers than in the one case in which Patria Potestas was undoubtedly present.

This brings us to make a further observation upon Sir Henry Maine's suggestion as to the connection between Patria Potestas and the tutelage of women. It assumes the prevalence of Patria Potestas. It can leave no impression unless it be assumed that Patria Potestas is known to have been at least of common occurrence—to have prevailed at the very least as commonly as the tutelage of women.

Now, to assume the prevalence of Patria Potestas was, no doubt, allowable in a writer who thought he had already sufficiently proved it. Having already proved this, he might legitimately try to show, by the evidence of resemblance, that Patria Potestas was the model after which guardianship had been moulded; and success in the attempt might even have strengthened the case for the prevalence of Patria Potestas. But if he is reduced to one instance of Patria Potestas, what becomes of the argument connecting Patria Potestas and guardianship? What, *à fortiori*, when it appears that the powers of guardians and the Patria Potestas have nowhere a real resemblance to each other, and that where the two undoubtedly co-existed, the law of guardianship, as we know it, exhibited no mark from which the magnitude of the Patria Potestas could possibly be conjectured? And here it has to be added that at Rome, where alone the two are known to have co-existed, the tutelage of women disappeared long before Patria Potestas.

When, putting aside assumptions, we inquire whether the pre-existence of Patria Potestas—that is, of the paternal power in the degree in which it was allowed by law in Rome—can be inferred from the tutelage of women, the following facts appear: (1) that Patria Potestas is known in one instance only, and that the tutelage of women occurred with it in a mild form, and, so far as is known, only in a mild form; (2) that there have been numerous examples of the tutelage of women occurring with lower forms of the paternal power, and that many of them show women under greater restraints than were known in Rome, although in Rome paternal power was at the highest; and (3) that in Rome, where alone the Patria Potestas and the tutelage of women appear together, Patria Potestas was the longer-lived. There is one fact more which will be mentioned immediately; but enough has already appeared to show that where systems of female guardianship are found without Patria Potestas there can be no inference from them of the pre-existence of a family system having Patria Potestas for its leading feature. The fact remaining to be mentioned, however, is of itself decisive of the question. It is that guardianship of women is by no means uncommon among peoples which acknowledge kinship through females only, among whom the head of the household, whatever his powers, is anything rather than a Paterfamilias, and whose family system is at all points in contrast with the family of the Patriarchal Theory.

Systems of female guardianship will undoubtedly help us in forming conclusions as to the condition of women in the regions in which they prevail. But it would seem that we should not be safe in making inferences from them as to the position and powers of fathers or even as to the system of kinship.

Before leaving this branch of our subject, it may be well to say, with reference to the history of systems of female guardianship, and by way of caution as to forming conclusions upon them, that, though Hindoo law-books contain passages which enlarge upon the dependence of women, or perhaps rather upon the propriety of keeping them ever dependent, there are ancient Hindoo facts which are of an entirely different tenor. It can be shown, for example, that, in early times, a Hindoo mother might have her children in ward. Even according to Manu, although the doctrine of the dependence of women is eloquently set forth in that work, it was improper for sons, their father being dead, to divide the family estate during the mother's lifetime. And reference has already been made* to an argument founded upon ancient authorities—an argument by a supporter of the Patriarchal Theory—designed to show that, in early times, among the Hindoos, after the father's death the mother was head of the family, and possessed, with the powers of control a father had, of the family estate.

It remains for us briefly to examine what Sir Henry Maine has said in *Ancient Law* of the relation of Patria Potestas to the slave.

There is a sense, he tells us,† in which an affirmative answer must certainly be given to the question whether the slave was in the early stages of society a recognised member of the family. "It is clear," he says, "from the testimony both of ancient law and of many primeval histories that the slave might under certain conditions be made the Heir, or Universal Successor, of the Master, and this significant faculty, as I shall explain

* At p. 58, foot-note. † *Ancient Law*, pp. 163–5.

in the chapter on Succession, implies that the government and representation of the Family might, in a particular state of circumstances, devolve on the bondman." "The Family," he proceeds, "consisted primarily of those who belonged to it by consanguinity and next of those who had been engrafted on it by adoption; but there was still a third class of persons who were only joined to it by common subjection to its head, and these were the Slaves." The tie which bound the slave to his master, he explains, was "regarded as one of the same general character with that which united every other member of the group to its chieftain;" and so the slave had, "in the beginnings of society, a definite place reserved to him in the empire of the Father."

In short, the slave might be his master's heir because he was among the persons subject to the Patria Potestas, and the relation of all such persons to their chief was substantially the same. And in the Patria Potestas we have the explanation of the slave's capacity of inheriting.

The Ashantees, we learn from Bowdich, acknowledged kinship through females only, and Bowdich, in his interesting essay on the constitution and laws of Ashantee,* speaks as follows of their law of succession:

"The most original feature of their law—that of succession—has been mentioned in the History, with the argument on which it is founded. It is universally binding. The course [of succession] is the brother, the sister's son, the son, the chief vassal or slave to the

* *Mission from Cape Coast Castle to Ashantee,* 1819, p. 254.

stool." The stool is found in every house, and is the emblem of authority, answering to throne. Bowdich goes on to state that in the Fantee country the principal slave succeeded to the exclusion of the son, who inherited only his mother's property, which was frequently considerable, and was inherited from her family independently of her husband.

Here we see that, with a system of kinship which is the very antithesis of Agnation, with a law of succession which preferred the brother and the sister's son to the son, there was room for the slave to be, under certain circumstances, the heir of the family. And, among the Fantees, he was heir while a son was actually excluded.

Nothing more seems necessary to show that the slave's capacity of being heir of his master does not imply the Patriarchal Family or "the empire of the Father," and that it has no special relation to Patria Potestas.

CHAPTER XIX.

CONCLUSION.

In one of the earlier chapters of this work an attempt was made to show that the evidence for the Patriarchal Family of Sir Henry Maine's theory, with its incidents of Patria Potestas and Agnation, should be exceedingly clear and strong.

The unfitness of his theory by itself to account for the actual forms of early societies, the apparent unfitness of the fiction with which he supplemented it to serve the purpose for which he supposed it to have been used, and the certainty that, if used, it had no effect whatever unless, as he imagined, it mysteriously—in a way admitted to be now unintelligible — enabled early tribes to impose upon themselves, appeared to lay the theory as a whole under the strongest suspicion—and would indeed have been enough to destroy any hypothesis which did not possess some powerful elements of popularity. It was suggested that, on account of the doubt surrounding the theory as a whole, good evidence might be demanded for the particulars involved in it. The complexity of the primeval family, as Sir Henry Maine has described it, which has made it, regarded as a primary social fact, a puzzle and a wonder at times to himself, appeared to be a still better reason why good evidence should be forthcoming of the prevalence of the family described, and a strong case be made out for believing it, if found, to have been primordial. It was suggested that the frequent occurrence of a family system founded upon kinship only through females supplied one reason more why the evidence for Sir Henry Maine's Patriarchal Family should be strong, and the reasons for holding it to have been primordial convincing.

After considering the evidence to which, in his various writings, Sir Henry Maine has referred us, the conclusion seems to be forced upon us that he has known the family which has the father's power in the degree of Patria Potestas, and in which Agnation gives the rule for kinship and inheritance, nowhere except in Rome. That this form of the family occurred in Rome appears then to be all there is of direct evidence to show that it was once universal. And that it occurred in Rome among a people very far from primitive is all there is of direct evidence to show that it was primeval as well as universal.

Against this direct evidence, however, we can scarcely help putting the evidence which investigations suggested by Sir Henry Maine's writings have disclosed to us as to the family systems of peoples known to us in a more primitive state than the Romans. But this need not be summed up here. It will suffice to recall that in the early Hebrew family we find neither Patria Potestas nor Agnation—and that, on the contrary, we seem to find beenah marriage and the relationships consistent therewith; while from not the Hebrews only, but from the Germans and from the early Greeks, we appear to get unquestionable indications of a system of kinship only through women having preceded the acknowledgment of kinship through the father. It need scarcely be said that the facts here referred to are evidence against the Patriarchal Theory in every form.

Of indirect evidence for the Patriarchal Family of Sir Henry Maine's theory there is less than of direct. For, putting Rome aside, Sir Henry Maine has never adduced a single good example of either Patria Potestas or Agnation.

That Patria Potestas was rarely met with had always been conceded. But though we have come upon old forms of the family in which the father's rights and authority were placed high—and, indeed, these tend to be great wherever fathers are recognised, and fill the father's place as protector of the family—in not one of them had the father the unlimited powers of the Paterfamilias. In none of the cases we have had to examine have we found him endowed by law or custom with powers of life and death and sale, the monopolist of family property, the only person in the family capable of having rights.

As to Agnation, it might be enough to say that

Agnation excluded from relationship, and especially from inheritance, the descendants of the women of a family; and that Sir Henry Maine has never specified any people other than the Romans who allowed no rights of inheritance to the descendants of women. This may be taken as showing sufficiently that cases of Agnation are extremely rare. A preference, more or less, for males in succession has been common enough. But that is not Agnation.

Even had cases of Agnation been numerous, however, Sir Henry Maine appears to have completely failed in trying to connect Patria Potestas with Agnation even by a hypothesis—so that cases of Agnation, if he had them, could not be used by him for recovering lost cases of Patria Potestas.

There seems then to be no indirect evidence for Patria Potestas or for the Patriarchal Family of Sir Henry Maine's theory to be had by means of Agnation. And in the preceding chapter it has appeared that there is really no other source of indirect evidence for them.

Of indirect evidence against the Patriarchal Family of the theory which has come in our way, it seems necessary to mention that given by the Levirate. An attempt to attenuate the prevalence of the Levirate among a particular people, and then to show how natural it was that priests or lawyers, as well as the multitude, should take a polyandrous view of marriage, and think the Levirate might be tolerated, could not possibly, in the case of so wide-spread an institution, be of any use in showing the institution to be consistent with a primeval and universal practice of monandry. And, if the Levirate must be taken to be of polyandrous origin, we have in it again strong evidence against the Patriarchal Theory, not only as Sir Henry Maine has stated it, but in every form.

The conclusion we are brought to is that, besides the occurrence of Patria Potestas and Agnation in the Roman family within the historic period, there is really no evidence to show that the Patriarchal Family, as Sir Henry Maine has described it, was primeval and universal.

It need only be added that the failure of Sir Henry Maine's theory to connect Patria Potestas with Agnation destroys its consistency as a theory. Unless some constant relation between these two can be made out, there is an absence of reason for their appearing together in the Patriarchal Family, or in the family of any period—even were there evidence to show that in particular cases they did so appear. And, at the same time, that failure leaves relationship unaccounted for.

So far as to the merely critical side of the present essay—which has dealt with an attempt to show that relationship began in the family, and was determined by a force operating therein. If, however, the reader has given his assent to the arguments by which that attempt seems to be condemned as a failure, he will also, it is hoped, be ready to admit that the result is not merely negative. For it has appeared at all points, not only that the phenomena dealt with are not intelligible on the Patriarchal Theory, but that they carry us back to a stage of society prior to the form of the family which has a father at its head, to the stage of polyandry and to the form of the family founded upon kinship through women only. The argument, therefore, has been throughout constructive as well as critical. And no slight part of the work is purely constructive. The value of the construction must be left to the judgment of the reader.

THE END.

CHARLES DICKENS AND EVANS, CRYSTAL PALACE PRESS.

Made in the USA
Monee, IL
04 November 2026

67387297R00207